T0225107

ROUTLEDGE LIBRARY EDITIONS:
ARTIFICIAL INTELLIGENCE

Volume 2

THE QUESTION OF
ARTIFICIAL INTELLIGENCE

THE QUESTION OF ARTIFICIAL INTELLIGENCE

Philosophical and Sociological Perspectives

Edited by
BRIAN P. BLOOMFIELD

Routledge
Taylor & Francis Group

LONDON AND NEW YORK

First published in 1987 by Croom Helm

This edition first published in 2018
by Routledge
2 Park Square, Milton Park, Abingdon, Oxon OX14 4RN

and by Routledge
711 Third Avenue, New York, NY 10017

Routledge is an imprint of the Taylor & Francis Group, an informa business

© 1987 Brian Bloomfield

All rights reserved. No part of this book may be reprinted or reproduced or utilised in any form or by any electronic, mechanical, or other means, now known or hereafter invented, including photocopying and recording, or in any information storage or retrieval system, without permission in writing from the publishers.

Trademark notice: Product or corporate names may be trademarks or registered trademarks, and are used only for identification and explanation without intent to infringe.

British Library Cataloguing in Publication Data
A catalogue record for this book is available from the British Library

ISBN: 978-0-8153-8566-0 (Set)
ISBN: 978-0-429-49236-5 (Set) (ebk)
ISBN: 978-1-138-58531-7 (Volume 2) (hbk)
ISBN: 978-1-138-58534-8 (Volume 2) (pbk)
ISBN: 978-0-429-50533-1 (Volume 2) (ebk)

Publisher's Note
The publisher has gone to great lengths to ensure the quality of this reprint but points out that some imperfections in the original copies may be apparent.

Disclaimer
The publisher has made every effort to trace copyright holders and would welcome correspondence from those they have been unable to trace.

THE QUESTION
OF ARTIFICIAL
INTELLIGENCE

Philosophical and Sociological Perspectives

Edited by
BRIAN P. BLOOMFIELD

CROOM HELM
London • New York • Sydney

© 1987 Brian Bloomfield
Croom Helm Ltd, Provident House, Burrell Row,
Beckenham, Kent BR3 1AT
Croom Helm Australia, 44-50 Waterloo Road,
North Ryde, 2113, New South Wales

Published in the USA by
Croom Helm
in association with Methuen, Inc.
29 West 35th Street
New York, NY 10001

British Library Cataloguing in Publication Data

The Question of artificial intelligence:
   philosophical and sociological perspectives.
   1. Artificial intelligence
   I. Bloomfield, Brian P.
   006.3   Q335
   ISBN 0-7099-3957-4

Library of Congress Cataloging in Publication Data

ISBN 0-7099-3957-4

Printed and bound in Great Britain by Mackays of Chatham Ltd, Kent

Contents

Acknowledgement

The contribution by James Fleck, 'Development and establishment in Artificial Intelligence', originally appeared in N. Elias, H. Martins and R. Whitley (eds), *Scientific establishments and hierarchies*, Sociology of the Sciences, vol. VI, 1982 pp. 169-217.
Copyright © 1982 by D. Reidel Publishing Company. It appears here with the permission of the publishers.

Figures and Table

Preface

Before the computer, the animals, mortal though not sentient, seemed our nearest neighbors in the known universe. Computers, with their interactivity, their psychology, with whatever fragments of intelligence they have, now bid for this place. (. . .) Where we once were rational animals, now we are feeling computers, emotional machines (Turkle 1984: 326).

Until computers and robots become as common as cars and until people are able to program and use them as easily as they now drive cars we are likely to continue to suffer from a certain mythological conception of digital computers (Searle 1982: 6).

Amid all the technological marvels of the modern world, computers must surely rank as one of the most powerful in stirring people's imaginations. Though the history of the electronic computer covers but a few decades, its rise to fame in the technological firmament has been truly meteoric; and while governments and industrial corporations perceive a strategically important role for computers within national economies, home computing has been added to the list of essential consumer devices along with the microwave cooker and deep freezer. Computing and the related technologies gathered together under the umbrella of 'information technology' are seen as the key to future prosperity, international standing, and economic power. In short, so we are told, modern societies are in the throes of an information revolution which will far outdo the nineteenth-century industrial revolution in its scope and impact on society. Indeed, information is actually becoming a commodity and the fear is that countries which do not seek to keep up in the information technology race will be reduced to the status of underdeveloped nations.

The dramatic, if not phenomenal changes in the size and power of computers furnish popular symbols of the seemingly relentless onward progression of computing. Thus, for example, the computers built in the 1940s to assist in the development and use of military systems occupied the space of a large room,

ix

whereas today, computers are considerably smaller and, para-
doxically, enormously more powerful. However, perhaps a
more relevant change in the development of computing is not
directly connected to questions of size or processing power so
much as to shifts in the pattern of computer usage. This has
changed almost beyond recognition: for example, personal
computers in the home are used to run software ranging from
programs for family budgeting to self-psychoanalysis, from
games of entertainment — such as the ubiquitous 'space
invaders' — to programs for educational tuition. Thus, from the
home to the school, from the factory shopfloor to the
management boardroom, from the hospital to the social
security office, it seems that no corner of society has been left
untouched.

Within this rising tide of computing one particular area —
namely, the science of Artificial Intelligence (AI) — is seen as a
key source of innovation and impetus for facilitating the
revolution in information technology. More specifically, there
exists the dream of so-called '5th Generation' computing —
initially announced by the Japanese MITI organisation in 1981,
but now pursued in many countries — wherein the hope is that
developments in AI will enable the construction of computing
machines which, for the first time, will have serious intellectual
abilities (i.e. *real* artificial intelligence). These computers will
have access to powerful knowledge bases encapsulating many
areas of expertise from which they will make complex inferences
(both inductive and deductive) in order to aid decision-makers;
and they will have the ability to communicate the results to
ordinary users in natural language.

Given the far-reaching nature of the potential ramifications
of 'intelligent' computers, AI is obviously an area of scientific
and technological endeavour which merits close inspection.
Indeed, in recent years there has been a large growth in the
number of published books and articles dealing with one or
other aspect of AI: as James Fleck has remarked, the subject of
thinking machines is one on which almost everyone has
something to say. With this burgeoning of the literature in
mind, the emphasis underpinning the present collection of
essays is very much aimed at bringing new perspectives to bear
on the subject. In particular, it is felt that recent work in the
philosophy and sociology of science provides a most useful
terrain on which to forge debate on some important, though

thus far little-discussed, aspects of AI. Thus, for instance, while a great deal has been written about the social impact of AI, there has been a scarcity of work dealing with the questions that AI provokes in relation to issues of sociological theory and method. Yet the prospect of non-social 'intelligent' machines communicating with each other and exchanging knowledge obviously challenges some of the tenets of sociology, if not its very autonomy. Similarly, there has been little discussion concerned with the actual social construction of AI, i.e. as an example of scientific culture to be studied in its own right. (Notable exceptions are: de Mey, 1982; Gilbert & Heath, 1985; Solomonides & Levidow, 1985; Woolgar, 1985.) In addition, where there have been philosophical critiques of AI, these have often been negative rather than constructive in orientation.

This collection of essays constitutes the bringing together of the work of scholars who have been pursuing research in the philosophy/sociology of AI for some time. While the sociologist of science is concerned with the social and cultural influences on the development of scientific knowledge, the philosopher is concerned with its validity or intelligibility. However, the two approaches can complement each other: because the sociologist of science wants to understand how social factors shape the *internal content* of scientific knowledge (as opposed to confining investigation to factors influencing the location, pace of development, or social impact of science), any philosophical elucidation of that content provides better material for socio-logical theorising. Though some of the arguments contained herein will inevitably conflict with some of the cherished hopes and convictions of the AI community, they should not be construed as negative *per se*. Each essay aims to increase the understanding of AI as *it is* — as opposed to how exponents of AI might *wish* it to be — and so this collection is actually constructive in outlook.

Although each contribution has been written as a self-contained essay, a number of common themes run through them and these are given different specific elaborations within the context of each argument. Chapters Two through Five, for example, are broadly concerned with the social/cultural factors which have shaped the development of AI; but while Chapter Two contains a rather more extended discussion of the theoretical background underpinning the sociology of AI, the other chapters present more in the way of detailed empirical material. (Thus, readers who are unfamiliar with the sociology of science might find the opening theoretical discussion of Chapter Two useful as an introduction to some of the later chapters.) Furthermore, while Chapter Two is concerned with the culture of AI in its widest sense (encompassing both academic and popular forms), the other contributions are concerned with the specific academic culture of AI, the various strands of its academic ancestry (e.g. neurophysiology, logic, mathematics, etc.), or with its sub-groupings — such as those concerned with the construction of expert systems, or the application of logic programming to law.

In Chapter One, Stuart Shanker presents a philosophical critique of AI and in particular of the 'Mechanist Thesis' that a machine (i.e. the computer) is (potentially) capable of thought. Sensitive to the charge, often made by exponents of AI, that philosophers are committed to outmoded beliefs about the human mind — that they are, in a Kuhnian sense, paradigm reactionaries — Shanker aims to make his critique a constructive one. He accepts that the philosophy of mind is under threat from AI, especially in its guise of cognitive science, but argues that (this disciplinary interest aside) it can benefit from philosophical clarification. He points out that historically, philosophy has often played a role in the development of science and that in this regard AI is no exception. For instance Shanker perceives an implicit scientism within AI which he considers to be a 'misguided attempt to pursue empirical solutions to purely philosophical problems' and he aims to prosecute a philosophical/conceptual clarification of such 'confusions' and 'futile goals'. Using the tools of logical grammar, he dissects the discourse of AI, and the features thereby revealed lead him to unravel the circularities in the arguments which form the basis of the Mechanist Thesis, as well as the ways in which they implicitly transgress the logical barriers between separate domains (e.g. the collapse of the qualitative question about the *content* of the information in a message, into the quantitative question of *how much* information it carries).

In Chapter Two Bloomfield sets out to discuss the culture of AI. The argument here departs from conventional socially/ culturally orientated treatments which (as already indicated) reduce either to questions of the deterministic impact of computers on society, or to the implications of thinking machines for the development of knowledge and culture in a more general sense. (Bloomfield argues that the premature foreclosure on the scope of sociological enquiry can be traced not only to previous traditions in philosophy of science but also to the theoretical outlook and assumptions which underpin the AI paradigm.) Instead, he takes the view that the sociologist must also consider the scientific culture of the conveyors of AI — that like other products of science, AI is a social construction. Borrowing from the work of Ludwik Fleck, Bloomfield considers the nature of AI as a 'worldview' or 'thought style' shared among a specific 'thought collective' which includes an esoteric 'elite' as well as a popular exoteric following — the 'masses'. The subsequent argument reveals the context-dependency of the beliefs of AI and the manner in which they are reinforced by the internal and external relationships of the thought collective. Thus, for example, the charge of 'paradigm reactionary' that is often levelled against AI's opponents is seen as a rhetorical resource by which the elite of AI seek to explain (away) hostile criticism and thereby legitimate their own position. During the course of his discussion, Bloomfield also compares and contrasts two of the more well-known evaluations of AI culture — namely, those of Turkle (1984) and Weizenbaum (1976); the questions asked here include the nature of the relationship (if any) between compulsive programmers ('hackers') and the academics of AI, as well as those which arise in respect of the argument that AI offers a less anthropomorphic and therefore more detached view of the human mind.

The third contribution contains a reprint of James Fleck's paper on the development and establishment of AI as a scientific discipline. It was one of the first to bring the tools of the sociology of science to bear on the subject of AI; originally published in 1982, it appears here with a postscript written to take account of the changes that have occurred in recent years. (In particular, Fleck refers to the tremendous commercialisation that has taken place within AI during the 1980s.) He examines the role played by scientific establishments in the USA and the UK during the development of AI, and charts the movement of key personnel among the major AI research centres. Among the other features discussed, Fleck lays emphasis on the processes of competition — both between different AI groupings (e.g. in conflict over the definition of 'AI') and between these and other more established disciplines

— to secure a monopolisation over the resources made available by funding agencies; and, in broader academic and cultural terms, to monopolise the 'means of orientation' through which we think and speak about matters concerning the human mind. A theme which runs throughout Fleck's argument is the interconnection between the internal and external social/institutional contexts within which AI became established (of which competition is but one feature), and its paradigmatic structure — including the inherent differentiation of research goals, the craft nature, and interdisciplinarity of the area.

In Chapter Four, Joop Schopman presents a detailed analysis of the so-called 'cognitive revolution' which, in the official history of AI, dates from the time of the Dartmouth Conference in 1956. This 'revolution' signalled the movement away from the 'hardware metaphor', with its emphasis on the use of the physical aspects of the digital computer as a model for understanding human intelligent behaviour, toward the symbolic information processing model — the 'software metaphor'. Schopman argues that the cognitive 'revolution' was certainly not a revolution in the Kuhnian sense but, rather, a stage in an ongoing process of change which was born in part out of the choices (among various possibilities) made by the leading actors. The themes he addresses therefore include the elements of continuity *and* discontinuity which straddle the putative revolutionary divide. On the one hand, for example, Schopman shows how a positivist heritage (stemming from the *Zeitgeist* of the preceding era when behaviourism had been dominant) has continued to exert its influence on AI up to the present day — even though the cognitive revolution had arisen in part from an acceptance of the inappropriateness of behaviourism for understanding human intelligent behaviour. On the other hand, he argues that some of the useful elements from the pre-1956 era became lost in the transition and, moreover (at least in areas such as robot vision), might have to be re-learned.

In Chapter Five there is a discussion of an applied area of AI — namely, the field of logic programming and its application to law. Here, Philip Leith (who also draws inspiration from Ludwik Fleck) explores the historical continuity between the sixteenth-century logic/law thesis of Ramism and the modern day version represented by Kowalski *et al.* which is centred around PROLOG — the logic programming language. Leith's discussion not only has something to say about the growth of this particular AI specialism — with its intellectual imperialism and virtual neglect of the major work, developed over decades, by philosophers of law — but also has more general import for the development of computer science and the nature of programming languages. He argues that the design of programming languages is not determined by purely technical matters (such as the speed of compilation or execution) but is also moulded by social features. Leith contends that programming languages often represent our 'involved' — that is, particular — views of how the world actually is (e.g. the belief that it is logically ordered); that in this regard the application of PROLOG to the law has been no exception; and that the future development of AI and computer science therefore depends on its adherents becoming less 'involved' and more 'detached' in their programming.

Finally, in Chapter Six Harry Collins addresses another applied area of AI, that of expert systems. Employing insights derived from Wittgenstein as well as his own previous work in the sociology of scientific knowledge, Collins seeks to illuminate some of the difficulties currently faced by knowledge engineers when building expert systems. These are not the usual *technical* issues which often tend to centre on choices between different knowledge representation formalisms or between different inference engines but, rather, the more fundamental problems that stem from the fact that knowledge is primarily *social* in character. Much of human knowledge, for example, is held on a tacit basis; it forms part of the background of shared assumptions on which culture is founded and as such cannot be articulated. This is one reason why human experts find it so difficult to verbalise their knowledge when being interviewed by knowledge engineers. Collins presents a model of culture which he uses to bridge the gap between the world of the knowledge engineer and that of philosophers/sociologists, and from this he develops a typology of expert systems and assesses the likely utility of the different types. He concludes with a discussion of some of the reasons why intelligent machines might come to appear capable of thinking like we do as humans, and how this 'illusion' might be dispelled.

REFERENCES

Gilbert, G.N. and Heath, C. (eds) (1985) *Social action and Artificial Intelligence*. Gower, Aldershot

de Mey, M. (1982) *The cognitive paradigm*. Reidel, Dordrecht
Searle, J. (1982) 'The myth of the computer', *The New York Review of Books*, 29 April, 3-6
Solomonides, T. and Levidow, L. (eds) (1985) *Compulsive technology*. Free Association Books, London
Turkle, S. (1984) *The second self*. Granada, London
Wiezenbaum, J. (1976) *Computer power and human reason*. W.H. Freeman, San Francisco
Woolgar, S. (1985) 'Why not a sociology of machines? The case of sociology and Artificial Intelligence'. *Sociology* 19(4), 557-72

Brian P. Bloomfield
January 1987

1

AI at the Crossroads

S.G. Shanker

Suppose I wished to show how very misleading the expressions of Cantor are. You ask: 'What do you mean, it is misleading? Where does it lead you to?'

Wittgenstein, *Lectures on Aesthetics*

1.1 PARADIGM REACTION(ARIE)S

It would be difficult to say which is the more striking: the pace of the internal development of Artificial Intelligence (hereafter AI), or its cultural assimilation. Rarely has a revolutionary technology matured so rapidly; and rarely have so many enjoyed the opportunity and the competence to discuss it. From the seeds of an obscure paper on the Decision Problem which appeared in 1936 and a pair of technical papers inaugurating Information Theory twelve years later, there have erupted thriving industries spanning the full spectrum of post-industrial society, providing gainful employment not just for computer scientists and communications engineers, but on the basis of their achievements, for biologists, neurophysiologists, psychologists, linguists, sociologists, and philosophers. Especially for philosophers.

Many from these professions are distressed by the Promethean implications of AI; only philosophers have been preoccupied with its intelligibility. Unfortunately, their efforts to clarify the coherence of the notion of 'artificial intelligence' are in danger of being smothered by the good intentions of their humanist cohorts. For it seems perfectly obvious to everyone concerned with the legitimacy of AI why philosophers should have devoted so much attention to what Ayer dismisses as 'the

1

sterile question whether machines can be said to think.'[1] The lingering influence of the Great Chain of Being manifests its presence in the pervasive resistance to any developments which might impugn mankind's privileged cognitive position on the evolutionary scale. In addition, since philosophers have hitherto enjoyed a 'monopoly' on questions concerning intelligence and the nature of mind, they are predictably reluctant to abandon one of the last remaining bastions of their authority without a prolonged struggle. From a sociological point of view the widespread philosophical opposition to AI was thus entirely to be expected: 'Such a reaction is not at all surprising given the sensitivity of such a goal to peoples' images of themselves'. For

> The AI approach is seeking to establish and legitimate a view of intelligence and the nature of mind which challenges the received commonsense view of mind and intelligence as something rather special and certainly well beyond the reach of scientific analysis.[2]

Professional interests apart, the AI-scientist still finds it puzzling how anyone who was familiar with the inexorable erosion of philosophy's sphere of influence at the hands of science could nonetheless oppose on *philosophical* grounds the exciting 'cognitive space' which has been opened up by AI. Is it simply because philosophers congenitally/professionally lack the imaginative flexibility to respond to new scientific paradigms? To philosophers clearly goes the sorry distinction of upholding the traditions of armchair science. Throughout their checkered scientific history they have emerged as the standard-bearers of archaic theories:

> Those unwilling or unable to accommodate their work to [a paradigm revolution] must proceed in isolation or attach themselves to some other group. Historically, they have often simply stayed in the departments of philosophy from which so many of the special sciences have been spawned.[3]

Contemporary philosophers are thus condemned to bear in silence the ignominy of the past failures of their forbears; what AI scientist is going to believe that they are not guilty of the same purblindness today? Or perhaps it is simply imperiousness?

Like an idealistic Jean Monet, the AI scientist would like nothing better than to see 'the distinctions among AI, psychology, and even philosophy of mind . . . melt away'.[4] However, philosophers steadfastly refuse to join in this common market; perhaps it is no surprise that so much of the philosophical opposition to AI today stems from Britain.

This is not simply a case of the analytic sons inheriting the sins of their natural philosophical fathers; for the picture developed by one of the most influential of their own peers seems to guarantee that the more they struggle the more they undermine their own efforts, however subtle these might become. The philosophical controversy which the advent of AI has unleashed is itself the crowning proof for the advocates of the Mechanist Thesis that theirs is a bona fide scientific revolution as Kuhn has described it.[5] For if philosophers are unable — or unwilling — to concede the significance as opposed to the feasibility of the Mechanist Thesis, this can only be because they are committed to an outmoded set of 'preconceptions' about the nature of man and machine. Viewed from an Enlightenment perspective of man's inalienable rational superiority, the new AI-paradigm will indeed seem unintelligible. Thus, the argument that '"AI" is not an analogy, it is a confusion' is exactly how it would appear to someone operating within the parameters of an outmoded paradigm; and the more carefully philosophers demonstrate the 'category mistakes' involved in the Mechanist Metaphor the more they confirm the naïve sociological thesis that they are paradigm reactionaries. On this picture the philosopher's greatest weapon — the demonstration that the Mechanist Thesis violates the rules of logical grammar — is rendered totally innocuous, and philosophical critique is silenced before it can even begin.

The AI scientist impatiently protests that the philosopher's objections to the intelligibility of the Mechanist Metaphor should apply to all scientific analogies; for, as many have pointed out, the very essence of scientific metaphor is its illogicality.[6] And yet philosophers are understandably loathe to condemn scientific analogies *tout court*. But then, why should 'artificial intelligence' be any different from (e.g.) the hydrodynamic analogy in the theory of electricity? At least, how can the philosopher object to the former but not the latter on *a priori* grounds? To the Mechanist what philosophers fail to appreciate is simply that it is always pointless to attack the

3

logical consistency of a metaphor; all that matters is whether the field in question displays its own internal logic which — as in the case of AI — has proven to be demonstrably fruitful. Even if current theories are wrong, that in no way entails that they are unintelligible. Hence 'Artificial Intelligence is neither preposterous nor inevitable: rather, it is based on a powerful idea, which very well might be right (or right in some respects) and just as well might not.'[7] The ultimate answer to what is seen as philosophical casuistry is thus: how do you reconcile your pedantry with the growing success of expert systems? That is, how could a *confusion* be so useful? The very success of the Mechanist Metaphor serves at one and the same time as compelling testimony to the fact that 'artificial intelligence' marks the emergence of a new paradigm, and *per consequens*, one which is only criticisable from a position *within* that paradigm!

If philosophers genuinely wish to participate in the revolution would they not be wiser to accept their officially sanctioned role as scientific underlabourers whose minor responsibility may be to clarify what these inferential machines can and will be able to do and how they accomplish this, but whose major challenge is to accept the light which this sheds on such perennial philosophical issues as the nature of the obscure processes which occur in either the mind and/or the brain? Thus Ayer indicates that the real problem here is not so much whether computers think as whether thinkers — i.e. their brains — compute.[8] This is very much the central theme of J. David Bolter's *Turing's Man*: 'By promising (or threatening) to replace man, the computer is giving us a new definition of man, as an "information processor," and of nature, as "information to be processed."' [9] The source of Anthony Kenny's qualms over the so-called 'homunculus fallacy'[10] might forseeably be removed by an arresting conceptual change brought about by scientific advance: 'Men and women of the electronic age, with their desire to sweep along in the direction of technical change, are more sanguine than ever about becoming one with their electronic homunculus.'[11] In the face of such dramatic socio-economic issues, philosophical squabbles over the cogency of 'artificial intelligence' may well seem a sterile controversy sparked off by technological impotence. However, what if the former only seem pressing because the latter have not yet been resolved? Or more problematically still, what if Ayer's two questions are merely different sides of the same problem — mutually reinforcing or pernicious, as the case may be — whilst humanist anxieties and mechanist fantasies alike are really the consequence of deep-seated philosophical confusions?

How can philosophers hope to break out of this impasse and escape the presumption that all we are concerned with here is a conflict between opposing *Weltanschauungen*? How can they attack the AI paradigm without thereby endangering the role of paradigms *per se*, or seeing their arguments rebuffed as yet another example of philosophy's ongoing countermarch from progress? And most significant of all, perhaps, how can they attack the Mechanist Metaphor without resembling Victorian parents forbidding their children to read novels because they are not 'true'? The most obvious starting-point is to reconsider what we mean by describing AI as a 'paradigm': an exercise greatly complicated, not just by the multiplicity of meanings which the term enjoys, but even more importantly, by the categorial distinctions which these exhibit.[12] For it is perfectly conceivable to speak of the value of the 'AI paradigm' from a sociological point of view — e.g. in terms of identifying the goals and/or techniques guiding a community of practitioners — and even to acknowledge the many creative possibilities suggested by the Mechanist Metaphor, while maintaining that the ideological cornerstone of AI — the Mechanist Thesis — is strictly unintelligible. The net result of referring to 'artificial intelligence' as either a 'paradigm' or a 'metaphor' is that this shelters the argument from philosophical scrutiny; for philosophy can have no conceptual quarrel with the evolution of new perceptual tools.[13] Even Douglas Berggren's argument that 'myth' results when metaphors are taken literally seems to disarm philosophical scrutiny *ab initio*.[14] However, no reputable science can hope to shelter forever behind the protection afforded by the analogical, and for this reason AI was forced — in order to acquire scientific credibility — to venture into the realm of theory, where the cries of metaphysics could no longer be dismissed as the symptoms of presuppositional confusion. For when philosophers oppose the Mechanist Thesis what they are objecting to is just that: a *thesis*, not a normative or metaphorical approach.

To revert to Kuhn, we can best appreciate what is problematic with accepting AI as a paradigm in the latter context by considering the type of problems to which this exposes us. Kuhn explains that 'one of the things a scientific community acquires with a paradigm is a criterion for choosing problems that, while the paradigm is taken for granted, can be assumed to have solutions.'[15] There is a tacit hint here — which several have picked up — that scientific hypotheses share with mathematical conjectures the fundamental trait that they can only be answered within an appropriate system. The proper philosophical response to this thesis, however, is that what scientific hypotheses share with mathematical conjectures is that they are only intelligible within an appropriate system, where rules exist for their use.[16] However, what concerns philosophy is neither whether these new problems are warranted — from whatever outlook — nor if they are unintelligible from some particular viewpoint, but simply, whether they are *answerable*: whether and in what sense they are *problems*. Kuhn, however, implicitly conflates this quintessentially philosophical interest with the former — sociological — concern when he continues: 'Other problems, including many that had previously been standard, are rejected as metaphysical, as the concern of another discipline, or sometimes as just too problematic to be worth the time.'[17] However, these latter remonstrations are by no means equivalent: to object that a paradigm is metaphysical is not at all the same as rejecting its utility. Indeed, in the sociological sense there are no grounds to say that a paradigm would be unintelligible when approached from another paradigm. On the contrary, in this respect the clash between paradigms would be perfectly meaningful to opposing practitioners; any dissension would be purely empirical: e.g. on pragmatic or aesthetic grounds.[18] To introduce the notion of *unintelligibility* the argument must shift to a grammatical plane where conflicting paradigms are comparable to, e.g. rival geometrical systems. (For example, it is not *false*, it is *meaningless* to assert in the context of Euclidean — as opposed to Bolyai-Lobatchevskian — geometry that the path of light rays might constitute a triangle the sum of whose angles is less than 180°.) But is philosophy committed to *any* — let alone a rival — conceptual scheme in the 'cognitive science of mind'?

The problem with Kuhn's picture is that it seems to force philosophy *nolens volens* into either a conventionalist or empiricist framework. One way or another, to challenge a new paradigm can only be to question its ability to account for diverse phenomena; and more importantly, to demonstrate one's prior commitment to an alternative — and in this case, supplanted — paradigm. The AI scientist would hardly be so foolhardy as to claim that computers will provide the ultimate paradigm for understanding the 'mysteries' of the mind or the brain: that such an analogy will adequately account for all mental or neurophysiological states. However, that does not entail that it is not a legitimate paradigm which suffices for a good many such facts; and hence, not subject to the philosophical objection that it is a confusion. On the contrary, it is clearly a valuable heuristic device, if only in focusing attention on which experiments to perform, etc. In response the philosopher asks: is it the Mechanist Metaphor which performs this role, or is this already in the process of becoming a dead figure of speech whose sole function is to enliven the prose discussions of cognitive/neurophysiological research?[19] However, even this question sounds suspiciously like a sociological matter; whether, and if so, why, so many sciences initially responded with such enthusiasm to the Mechanist Metaphor is, perhaps, an issue which lies irrevocably outside the compass of philosophy, whose sole concern is with the resolution of philosophical problems. Treated as a purely philosophical issue, however, this brings us to the point that what philosophers are rejecting is not a paradigm (analogy/metaphor/myth) in any normative sense: it is a spurious 'philosophical thesis', which as such demands conceptual clarification as opposed to empirical confirmation or epistemological refutation.[20]

Whereas AI strikes the sociologist as an exemplary case of 'paradigm change induced by a new paradigm',[21] to the philosopher it appears as a burgeoning science encumbered with prose interpretations which are little more than misguided attempts to pursue empirical solutions to purely philosophical problems. As Steve Woolgar points out, 'several significant philosophical pigeons are coming home to roost' in AI; for if 'AI turns out to be a feasible project, this would vindicate those philosophies which hold that human behaviour can be codified and reduced to formal, programmable and describable sequences.'[22] In other words, AI holds out the promise of providing a definitive solution to obdurate problems in the philosophy of mind (for the reference to 'human behaviour' here is a — regrettably biased — allusion to cognitive abilities). On this picture, all philosophical controversies only remain such because science has not yet developed the technical means to transform the abstract into the experiential. Leaving aside

7

the etymological changes which 'philosophy' (or 'natural philosophy') has undergone,[23] it is important to recognise that this condescending — i.e. scientistic — attitude towards the relationship between philosophy and science concentrates unduly on only one party to the felony; for there are the same number of discredited scientific theories which were dissolved as much by philosophical clarification as new discoveries (*pace* the history of phlogiston and spontaneous combustion, or perhaps a little closer to home, phrenology). Admittedly, these are now often disgraced by the name of 'metaphysics': but that in itself should serve as a reminder of the key role which philosophical critique has so often played in 'paradigm revolutions'.

The greatest irony in this whole issue would be if it is AI which is found guilty of the conceptual obsolescence which it would attribute to philosophy: i.e. if it is the advocate of the Mechanist Thesis who is the real 'paradigm reactionary'. For the long-standing ambition to construct an automaton and/or discover a harmless manifestation of the homunculus rests on a framework which not only licenses, but implicitly demands the total displacement of philosophy by the 'cognitive science of mind'.[24] At one end of this 'philosophical' (*sic*) scale stands behaviourism; at the other, mentalism. Despite their deep-rooted conflict over the grammar of psychology, however, both are committed to the fundamental premiss that the possibility or otherwise of constructing a 'cognitive machine' was a strictly epistemological affair. It is no coincidence that this picture has been refurbished in recent years on the basis of the thesis in the philosophy of mathematics that there is a species of undecidable — infinitary — propositions which outstrip the recognitional capacities of finite intellects: and thus, that Gödel's second incompleteness theorem should have come to play such an important role in the debate on the cogency of 'AI'. The central question in the latter was whether the full range of cognitive abilities displayed by the (human) mind could ever be artificially simulated. Hence, the emphasis in the interminable debate between mechanists and humanists was on *relative* capabilities: i.e. whether machines could ever be *as* creative (free, responsible, etc.) as minds.

The great innovation posed by contemporary versions of the Mechanist Thesis was simply to abandon the analogical; the Mechanist Metaphor is, in fact, little more than a faint reminder of earlier attempts to settle the issue by comparing the mind with whatever was the prevailing technology.[25] No matter how sophisticated the metaphor, it would always remain vulnerable to the 'anthropomorphic prejudices' dictated by 'human chauvinism'. The secret to the recent — doctrinal — successes which AI has enjoyed lies in the dramatic shift in focus which it instituted. The early pioneers of AI conformed to the established methodology: what, they asked their humanistic colleagues, would you deem the necessary and/or sufficient qualities of intelligence? The Turing Test was really just an example in the genre of thought-experiments which mathematicians since Poincaré have put to such impressive use. However, as Donald Michie and Rory Johnston complain in *The Creative Computer*, whenever an AI simulation displayed all of the requisite properties, beleaguered humanists would simply fall back on the dogmatic reply, when told 'how the program works': 'So that's all! I don't call that intelligent.'[26] Over the past decade AI has become far more subtle, however, and far more convincing. The strategy is no longer to imagine 'computer simulations' that would satisfy the most stringent of humanist demands; it is now to throw caution to the wind and elucidate human cognitive abilities as — literally — a species of computer-like operation.

Admittedly, AI still 'wants only the genuine article: *machines with minds*, in the full and literal sense'; but now the key to the realisation that 'This is not science fiction, but real science, based on a theoretical conception as deep as it is daring' lies in the premiss that 'we are, at root, *computers ourselves*'.[27] Or rather, that the human mind embodied in the brain and the computer are both sub-genera in the amorphous class of 'information-processing' systems. Of course, had this extended argument been confined to the realm of thought-experiment AI would not have transcended the metaphorical. However, to the great fortune of the struggling mechanist, a vanguard of leading psychologists and neurophysiologists took up the cause. When Richard Gregory remarks at the beginning of *Eye and Brain* that, 'Like a computer, the brain accepts information, and makes decisions according to the available information',[28] or J.Z. Young considers at the outset of *Programs of the Brain* 'how the organisation of the brain can be considered as the written script of the program of our lives',[29] they are not just interested in the heuristic value of a metaphor! (see Section 1.2) However, the argument demands more than rhetoric if it

9

is to carry any more weight than the Turing Test. Hence, the *Gedankenexperiment* has been superseded in cognitive psychology by, for example, the Ames Room or 'reaction-time studies': putative examples of the translation of early computationalist insights into something resembling empirical terms (see Section 1.4). Even more important, perhaps, is the fact that as a result of the latter developments AI can now hope to downgrade its high — philosophical — profile and assume a much more inconspicuous — and for that reason alone, more effective — role as merely one in the confederacy of interrelated sciences which together constitute the new 'meta' discipline of cognitive science. The only serious obstacle to the growing harmony of this union is philosophical intransigence.

There is an obvious historical — and perhaps internal — tension between science and philosophy; if the practitioners from each now view the activities of the other with ill-disguised mistrust, it is largely because both sides have so frequently overstepped the bounds of their rightful operations. The simple question which that leaves us with is therefore: which have we here, a case of philosophical primitivism or yet another of the prose confusions that fill the archives of pseudo-science? To be sure, AI — *qua* research activity organised around list processing programming languages — is not about to collapse because of a philosophical critique; but the Mechanist Thesis is doomed to suffer a grievous injury should it be the philosophical qualms which are vindicated. Before we can approach this vexed issue, however, the question still remains whether this purportedly distinctive philosophical concern with conceptual clarification really does differ, as was merely assumed above, from traditional humanist worries over what McCluhan memorably stigmatised as 'the numb stance of the technological idiot'. How does Wittgenstein's preoccupation with Bishop Butler's motto, 'everything is what it is, and not another thing', differ from David Bohm's ominous warning: 'As we perceive and talk, so we will think and act and, therefore, so we will be'?[30] Does not the very epigraph to this chapter belie the attempt to draw any sharp conceptual distinction?

The point here is that, although the philosopher's animus may — and probably does — proceed from humanist anxieties, the techniques which he adopts are *sui generis*, and the problems which he attacks are categorially distinct, if not prior. Where the humanist seeks to constrain the futurologist's technological raptures with realistic pictures of the type of people or society that would result if his/her visions should materialise, the philosopher's concern is with those unintelligible postulations which masquerade as hypotheses; and the tools he adopts to combat these phantasms are those of logical grammar, not the imagination. It is because AI has harboured the much more serious step from Mechanist Metaphor to Thesis, therefore, that the services of conceptual clarification are pressed into action. If AI were simply a paradigm we could rest content with the characteristically humanist interest in its cognitive impact: viz. on how it is influencing scientific perceptions and explanations of natural events. However, this already concedes the very premiss which most distresses philosophy: for if the question 'Do you not see where all this is leading?' is intended to limn the type of society and/or conception of humans that is likely to evolve if 'thinking machines' should acquire semantic as well as mechanical viability, you will merely have argued within Kuhnian terms; i.e. accepted the legitimacy of the Mechanist Thesis as a paradigm, but challenged its desirability on extra-paradigmatic grounds. Philosophy's target, however, is not the AI paradigm *per se* — as viewed, for example, in sociological terms — but the philosophical confusion which has inspired the very name of the field: and thence many of the futile goals which have hitherto impeded AI research. Humanists may find the prospect of 'Turing's man' repugnant, but it is nonetheless a prospect which, in all the blindness of his technological energies, they find *homo faber* capable of creating. To philosophers, therefore, goes the task of showing why 'Turing's man' is nothing of the sort.

At some point, however, philosophers must ask themselves why it is that, despite the meticulous conceptual care which they have exercised in their investigations into the logical grammar of normative or intentional *vis-à-vis* physiological or mechanical concepts, their arguments can have so little impact: that even an opening sentence can satisfy a scientist that all that follows belongs to the sphere of semantic posturing. It is always tempting to dismiss such hostility as the consequence of conflicting *Weltanschauungen*, and hence not susceptible to philosophical reason. However, such a fatalistic attitude itself wavers precipitously between the sociological and the hermeneutic; whereas what we want is a philosophical understanding of how it is that such vigorous philosophical scrutiny can be

11

rejected out of hand as nothing more than a species of technological primitivism. Our task in the following section, therefore, will be to consider whether perhaps the main reason why it is so easy for cognitive scientists to ignore Wittgensteinian intrusions into their fields is that, from their perspective, such exercises in conceptual clarification display an adherence to the very paradigms which their new models are intended to supersede. This is why they need a reminder of the aetiology of 'information-processing' *vis-à-vis* the logical grammar of cognitive abilities: for in arguing that there are different manifestations of 'information-processing systems' they contend that, given the Mechanist Thesis the Cognitive Thesis is unobjectionable, and given both together it obviously follows that there are different *kinds* of minds, ranging from the mechanical to the biological. Such a methodology is by no means foreign to the history of science; and for that very reason, neither is the present depth of philosophical involvement.

1.2 THE PHILOSOPHICAL 'WHY?'

The danger to the philosophy of mind posed by the gathering forces of cognitive science rests on an all-too-familiar tactic which, for all its claimed novelty, has nonetheless enabled the former to prepare its defences. However, in recent years a more subtle menace has emerged on the horizon, threatening to enclose philosophical investigations into the grammar of psychology in a scientific pincer action. In the classic 'philosophy versus science of mind' dispute, the concern with conceptual clarification remained the philosopher's exclusive preserve; but as can be gleaned from the preceding section, now even this is being encroached by the latest scientific aspirant, the sociology of science — particularly in light of the latter's growing interest in 'discourse analysis'. As Michael Mulkay presents the issue in *Science and the Sociology of Knowledge*, the very fact that the sociology of science is so recent a development — that science was for so long regarded as an exceptional epistemological case, as such situated outside the purview of the sociology of knowledge — was itself a result of long-standing 'philosophical preconceptions'. It was only because philosophers of science were forced, in the early 1960s, to venture into sociological waters in order to resolve the otherwise intractable 'traditional problems of their own disciplines' that this prejudice was finally exposed. 'Gradually, these new ideas have entered sociology, helping to undermine the epistemological assumptions which had virtually required the sociology of knowledge to treat science as a special case.'[31] Thus, both AI and the sociology of AI can seize on one and the same malign influence as a major source of the purely *a priori* obstacles inhibiting their initial development.

Whether or not such a sociological thesis can itself be substantiated, it is important to see how it perforce assumes a fundamental conceptual affinity between philosophical and sociological questions and their methods of solution. However, assuming that sociologists are not about to abandon their pursuit of scientific credibility, this could only be the case if, *contra philosophos mores*, the former problems should turn out to be intrinsically empirical. To be sure, both sociologists and (Wittgensteinian) philosophers of science see their activities as detached from immediate scientific concerns — as primarily descriptive, not prescriptive — and both share the same overall goal in their respective approaches: to 'strip away the formal side of science, and show what (is) *really* going on.'[32] Hence, both are particularly interested in the significance of scientific 'interpretation', and the manner in which this relates to the mechanics of theory-construction. Subtle differences in outlook remain, however, and although the two disciplines might begin with similar questions, their divergent purposes soon make their presence felt. The sociologist looks at scientific interpretation primarily in order to ascertain how social and cultural factors may have influenced the creation of scientific knowledge; but philosophers only scrutinise scientific 'prose' in order to determine whether an interpretation really qualifies as such, and if not, how it distorts the experimental results that were obtained.

From a sociological point of view that attitude may, unfortunately, be the source of much of philosophy's present ills; but philosophers must not allow any latent hostility here to blind them to the benefits to be afforded by a modest incursion into the areas opened up by the sociology of science. In fact, the latter highlights several crucial factors for our understanding of AI: the role of metaphor in the evolution of cognitive science, the cultural reasons why cognitive scientists have gravitated towards the mechanist and computationalist models, and the —

regrettably key — role which theorising philosophers have played in many of the assumptions that have proliferated. Most important of all, perhaps, is to recognise that, however glaring the philosophical problems might appear, beneath the metaphor there is indeed a revolutionary technology to be dealt with: that the 'gas' which has attended these developments is merely the product of a new science's struggle to find its bearings. As it stands the computationalist burden from which cognitive science suffers provides a paradigmatic illustration of the pernicious influence which preconceptions can exert. If the sociology of science — together with the history of ideas — can show us why it is that cognitive science has been inexorably drawn into the web of mechanist preconceptions, it will have performed an invaluable service to philosophy; for it is essential that we understand why the Mechanist and Cognitive Theses were first adopted and continue to win new converts if we are to succeed in our efforts to undermine the picture, thereby enabling the computer to become a potentially fruitful and illuminating analogy for ongoing developments in psychology and neurophysiology; and most important of all, to expedite AI's struggle to transcend the present phase in which it is so vulnerable to philosophical criticism.

Whereas the sociological 'Why?' places scientific interpretation in terms of its motives, influences, presuppositions and intuitions under the microscope, however, the philosophical interrogative must remain firmly focused on the proprieties and internal pressures of logical grammar. The sociology of science, in its post-Kuhnian vogue, is thus concerned with the dynamics of paradigm evolution; and philosophy, in its post-Wittgensteinian infancy, with the stringent demands of clarity and the intricacies of conceptual confusion. Ideally, the sociologist is an *amicus curiae*, standing apart and looking impartially at the trials s/he is reporting; but the philosopher cannot ask 'Why?' without actively engaging in a conflict *as a combatant*. (Whether or not s/he will be admitted is, of course, another matter; but even more important is the question whether the court's ruling that s/he can only participate as either plaintiff or defendant is upheld.) Both fields seek an overview — which by definition must be objective — therefore, but in the philosopher's case this will only succeed in its intended purpose if it not only clarifies, but when necessary, obviates the assumptions inspiring what philosophical scrutiny reveals to be a case of metaphysical or epistemological confusion: for unlike the sociological version, philosophical 'description gets its light, that is to say its purpose, from the philosophical problems' (Wittgenstein's *Philosophical Investigations*, Blackwell, 1953, § 109). Hence philosophy, we might say, does not suffer from but rather deliberately embraces a Heisenberg Effect; and the question which this raises is — from a sociological perspective at any rate — whether such a self-interested party is capable of discerning what is really going on in a paradigm revolution. However, as we shall come to see, where the sociology of science may be responsible for mapping the interstices between theory and fact — i.e. charting the extent to which any empirical statement is theory-laden — philosophy is more concerned with undermining such inter-dependencies if and when the occasion demands it. Thus, when a philosopher scrutinises a particular statement s/he is not challenging the utility of the underlying theoretical framework, but rather, the cogency of the expressions yielded by the theory: viz. their failure to function *as empirical propositions*.

In the welter of interrelated issues which have been raised by cognitive science, the Wittgensteinian philosopher's attention is riveted on two — logically separate — conceptual tracks. The preliminary task is to determine the 'logical geography' of mental, normative, or intentional versus neurophysiological, psychological/linguistic, communications, or computing concepts: to clarify the internal consistency and implications of each level, to ascertain whether, and if so how, they interact with one another, and to elucidate the confusions which result when the logical barriers between disparate domains are transgressed.[33] The sociologist, however, is more interested in how the 'socio-cognitive characteristics of AI . . . have been shaped by and in turn influence the history and development of the field.'[34] Sociologists and philosophers thus approach, not just AI, but each other's bearing on its development, with markedly different attitudes and objectives. If the sociologist in question does suffer from either scientist or humanist proclivities it is almost certain that s/he will attempt to persuade one of the growing number of participants in this expanding debate to moderate their demands for the sake of intellectual as well as professional harmony. Of course, the philosopher also wishes to curtail hostilities, but in this case that is all too often nothing more than a euphemism for vindicating his/her position.

Undoubtedly, their two approaches should feed off one another, but their seemingly inevitable estrangement once again turns on their disparate outlooks on conceptual developments. The former is necessarily eager to grasp why there is such pronounced conflict between cognitive scientists and their Wittgensteinian critics; for this in itself constitutes a key element in the sociological interpretation of the cultural factors influencing the development of AI. The latter worries that, by casting the conflict between cognitive science and philosophy in strictly sociological terms, there is a pronounced danger of sanctioning conceptual confusions under the all-embracing cover provided by the notion of 'paradigm-evolution'.

In particular, what most concerns us here is that where the sociologist looks at the manner in which the ties between AI and the neurophysiological, psychological, and psycholinguistic sciences have entrenched and orientated AI, the philosopher scrutinises the manner in which these bonds have illicitly nourished one another by further obscuring the assumptions on which each rests. In this respect it was slightly misleading to suggest in the preceding section that AI merely transferred its focus from the Mechanist to the Cognitive Thesis. The actual transition was rather more subtle: it was soon realised that the Mechanist Thesis would never be able to vanquish humanist polemics on the basis of its own merits — however sophisticated future computer simulations might become — and thus a move to 'hard science' was called for, where 'preconceptions' would be forced to do battle with objective facts. The somewhat abrupt and called-for shift was facilitated, however, by the concurrent development of psychophysiology: the attempt 'to find mathematical functions relating body and mind — or relating behaviour and consciousness — by mathematical descriptions.'[35] With one foot in the door of psychology and the other in that of neurophysiology the elimination of philosophy from the 'cognitive science of mind' could, so it seemed, be made effortlessly and — more importantly — irrevocably. Hence, whatever doors might in future open up from this initiative, none would lead back to naïve 'armchair speculation'. The major problem with this transitional step, however, was simply that it harboured — under the so-called psychological component — the very fallacies under consideration, which were immediately transmitted to the neurophysiological counterpart where they could be scientifically embellished before returning to do duty in their original psychological framework. It is precisely this interlocking network of mutually reinforcing assumptions that philosophical critiques of cognitivism are striving to extirpate.

If the Cognitive Thesis were an ordinary technological metaphor which — in Berggren's phrase — some were in danger of abusing, we would merely have to deal with the logico-grammatical anomalies created by comparing the so-called 'cognitive structure' of the neurophysical states of an organism with the various operations comprising a computer system. However, note how, for example, Jerry Fodor presented his theory that 'understanding a sentence (is) analogous to computational processes whose character we roughly comprehend.' In Fodor's own words, 'On this view, what happens when a person understands a sentence must be a translation process basically analogous to what happens when a machine "understands" (viz. compiles) a sentence in its programming language.'[36] His ensuing theory would not have been nearly so interesting had there been no mention of machine understanding, and the parentheses were removed from 'compiles'. Moreover, it is imperative to see just how much work is being performed by the grammatical barricade thus erected around 'understands': for the computationalist analogy only makes sense on the basis of the assumed credibility of the Mechanist Thesis. The success of the Cognitive Thesis would indeed ratify the claims of the Mechanist Thesis, but only because it had proceeded by assuming the cogency of the latter. That which was the source of the metaphor has thus become its chief beneficiary, in what begins to look suspiciously like a case of trans-disciplinary circularity. But then, such a manoeuvre would be highly reminiscent of the original attitudes to the categorial transgressions inspiring the Cognitive Thesis.

In his influential 'Minds and Machines' Hilary Putnam dutifully warned that we must always take pains to avoid category mistakes in our explanations of human behaviour: hence, we must be careful to distinguish between two levels of logical grammar involved in, for example, the description of avowals. However, he then immediately queried in what sense these physiological processes and intentional acts constitute the 'same event'. Not even inverted commas can distance this question from the resulting violation of the original emphasis on categorial disparity, for the tension on which the argument turns was only created by its initial assumption that there are two levels of logical grammar involved in the expression of avowals: that there is some sense in which 'Jones *stating* that he feels bad' is the *same event* as 'Jones's body producing such-and-such sound waves'.[37] However, the whole point of clarifying the *categorial* distinction between avowals and neurophysical states is that this prohibits the postulation of a psychoneural identity thesis, which by its very nature can only operate if it treats mental and physical reports as alternative descriptions of the same phenomenon — viz. 'human behaviour' — somewhere conjoined in the mysterious ether of 'events'. To be sure, an utterance — just like an object — can be described in two — categorially — different ways; having recorded such a demarcation, however, we must not then set about trying to dissolve this barrier, either by fusing the two levels of description together in the object of the description, or else by reading the original normative/intentional concepts back into the purportedly new level of explanation. On this approach, a categorial distinction is reduced to an explanatory hiatus, waiting to be bridged by a theoretical construct. Thus, the initial appearance of an insuperable conceptual barrier to such reductionist analysis must be removed — in this case, by tacitly conflating the expression of an avowal with the utterance of a sound.[38]

One way around the persevering objections that the cognitive abilities consequently attributed to the central nervous system were only the result of the category transgressions covertly built into the theory at the outset, however, involves a subtle variation on the classical theme of technical catachresis, in which it is argued that what are misleadingly referred to as 'vernacular' mental concepts are really primitive theoretical constructs which as such are not so much displaced as discarded by new theoretical concepts.[39] There can obviously be no category conflicts when one of the categories has been forcibly removed. Thus, as with the struggling 'theory of meaning', the computationalist must defend the premiss that our 'ordinary language' manifests an inchoate theory; in this case, of the mind. However, this proceeds from a similar motivation as the above approach — to remove the appearance of an insuperable conceptual barrier — and creates similar pressures to those noted above: here as a direct result of the initial assumption that, for example, avowals are — unbeknownst to 'ordinary' language speakers — really *referential* statements. From this premiss it is indeed possible to build up a complex picture of our 'primitive' theory of mind, originally inspired and to a limited extent remediable by philosophy: a far more promising task than clarifying the grammar of mental concepts, which can but sustain what by definition must now be treated as a defective theory. It is thus no coincidence that Dennett's theory should lead to similar consequences as Putnam's original approach; for both seek to establish that normative and intentional concepts can in some sense be attributed to the central nervous system in order to generate a comprehensive explanation of human behaviour. Hence, both assume the *feasibility* of that which remains to be proved; and with philosophical problems, overcoming the initial obstacles erected by logical grammar is everything.

So also we are told that we are presented with independent theories — the Mechanist and Cognitive Theses — when all we really have is one and the same idea in two different applications. Hence, these theses are *internally*, not externally related: in exposing the one we *pari passu* subvert the other. Needless to say, the cognitivist could not afford to allow such a line of attack to proceed unchecked (were s/he suddenly to become aware of its gravity). However, in response to the philosophical attacks on the category mistakes sustaining 'artificial intelligence' — in its mechanical/computational version — cognitivists can simply fall back on the notion of the *internal consistency* of a model/paradigm. With so much corroboration waiting to be found in cognitive writings, philosophical objections can be brusquely swept aside. For to object that a thesis is unintelligible because it violates the laws of logical grammar is obviously designed to convey that any 'questions' which this raises cannot be answered; and this seems irrefutably denied by the great advances realised by practitioners in the field, as measured by the spreading influence of the paradigm. Hence, the emphasis on type-category restrictions once again emerges as the common fixation evinced by paradigm reactionaries. For, of course, on either of the old categories, questions from each level would be quite unintelligible on the other. However, what is being forged here is some entirely *new* category, where the questions being asked can indeed be answered — but only on that new level!; and it is the gradual enlargement and refinement of the new category that is meant by the 'internal consistency of the paradigm'. Thus, to judge the intelligibility of any statement from a level other than that of the new paradigm is both loaded and itself categorially misguided.

The panlogicist fallacy sustained in this argument, however, is that it once again ignores *ab initio* the significance of category restrictions: for in distinguishing between two levels, our whole point is that these cannot be conflated: neither in reductionist nor expansionist terms. Nor can the pressures thereby created be released by ascending in Mannheimian fashion to some 'meta' theoretical level where categorial conflicts become second-order hypotheses. Hence, one reason why Wittgensteinian philosophers have so carefully and convincingly plotted the network of category mistakes suffusing computationalist psychophysiology is in order to clarify how the appearance of a new category is an illusion bred by the preliminary violations of logical grammar which induced the resulting theory. The freedom which science enjoys to create new concepts may be unbounded, but the constraints which it suffers to respect the logical grammar of existing concepts displays the unyielding 'hardness of the logical must'. Internal consistency alone is no guarantee of significance — as Lewis Carroll and Colin Turbayne in their different ways both amply demonstrated; and it is for this reason that the 'internal consistency of the paradigm' is by no means to be shunned by the philosopher. On the contrary, it is this which provides the basis for what is one of his most potent tools: the *reductio ad absurdum*. In fact, it is by following through the implications of the Mechanist/Cognitive Theses that the philosopher hopes to reveal the source of their mutual presuppositions. One great danger with this strategy, however — apart from the ever-present hazard of seeing a *reductio* misconstrued for a direct proof — is that it is so easy to confuse a *reductio* with an *argumentum ad absurdum;* i.e. to interpret the philosopher's intentions as trying to remove one theory in order to make room for his or her own.

As we saw in the preceding section, the greatest obstacle to understanding the nature and thus the import of philosophical criticism is the presumption that, *qua* participant in scientific dialogue, philosophy must operate within the parameters of the Kuhnian model of paradigm conflict, which provides a rationale of sorts for what are implicitly treated as 'untestable' assumptions. The paradigm model appears to shelter individual propositions from direct attack; only theories can be criticised, and these solely from the standpoint of a rival paradigm. As Mulkay explains,

> The idea that factual reports are formulated in terms of their associated theoretical or metaphysical presuppositions seems to imply that each theory can only be tested in its own terms, yet that within its own terms it is immune from refutation.[40]

Hence, the Kuhnian model readily assumes that philosophy confronts science *on the same footing*: that philosophical 'theories' compete with scientific, and that the philosopher cannot understand the AI paradigm for the simple reason that s/he is operating from an incompatible (one hesitates to use the term 'incommensurable' these days) conceptual framework. However, the very premiss of this picture is misguided; philosophy does not set out to subvert factual claims: it shows that no factual claims have been made! In other words, philosophy is not involved in the business of *refuting* a theory — neither in empirical nor conceptual terms — for that would still be to demand citizenship in the scientific community. Thus, it is misleading to argue that philosophy's paramount interest in science is epistemological: e.g. to demonstrate that the propositions of a given theory are unverifiable or unfalsifiable because they transcend our 'recognitional capacities'. Rather, it is to show that and why it is — logically — impossible to treat certain expressions *as propositions*: to scrutinise the logical grammar of what are all too readily accepted as meaningful albeit problematic propositions and conjectures.

'Paradigm consistency' clearly affords no protection from this kind of critique: for the fact that some expression may have been legitimately derived from a set of hypotheses by no means guarantees that expression is meaningful if the 'hypotheses' themselves or the picture which has inspired the interpretation of the theory is suspect. Herein lies the key to the philosophical 'Why?': what particularly interests the philosopher are the *conceptual pressures* which have compelled a theory to proceed in a certain direction; for example, why the Cognitive was such an inevitable consequence/complement to the Mechanist Thesis. The meeting-point between the two is a 'therefore': the former maintains that a system has performed such-and-such operations and therefore it understands, while the latter contends that a brain has processed such-and-such signals and therefore it

21

computes. It is a 'therefore' which assumes that there is a species of rule-following — either identical to, subordinate to, or independent of ordinary normative behaviour — which can be directly mapped between electro-mechanical and neural systems. Here also lies the meeting-point between AI and philosophy, given philosophy's special charter to reveal the conceptual sins committed by the unintelligible 'therefores' that breed in mathematics and science: the *prose* which, far from following from the preceding proof or evidence, conceals or demands a categorial leap to desired conclusion. Hence, this is not a case of ascertaining whether the facts *warrant* the conclusion: it is whether the conclusion is *internally related* to the facts. Finally, as was hinted at above, it is here where the sociology of science and philosophy must meet. It is obviously not enough to list the category mistakes which might undermine some particular interpretation and then dismiss the underlying facts as *per consequens* of no conceivable empirical interest when, as was stressed at the outset of this section, philosophy's ultimate goal is to show 'what is really going on'. The removal of some prose confusion is merely the first step towards 'the consummation of philosophy': a 'work which does not cheat and where the confusions have been cleared up.'[41]

Just as there can be no terminal *a priori* objection to an analogy — if only because a metaphor can be interpreted in myriad ways — so also a thesis cannot hope to lurk indefinitely beneath its metaphorical cloak. Hence the philosophical quarrel with AI is not with the Mechanist Metaphor *per se*, but rather, with the theses which emerge when the metaphor is stripped away. Indeed, from another point of view — the so-called 'negative analogies of the model' — the computer metaphor might become a valuable aid to psychological or neurophysiological research; e.g. if it were used to highlight some of the lacunae in each or the perplexities which inevitably result from category transgressions induced by mechanist preconceptions. Certainly there are superficial grounds for such an analogy. It is difficult for the cognitive scientist to see, for example, how anything other than an over-zealous desire to deprive the computationalist theory of even the slightest latitude could lead one to deny that there is an obvious parallel between the electronic processing of data in a CPU and what Putnam describes as the 'digital elements' of the brain: 'the yes-no firing of the neurons'.[42] However, the trouble with this premiss is, of course, that it seems to provide succour for the computationalist theory of memory, in so far as this 'correspondence' only qualifies as such on the assumption that what we are comparing are similar methods of transmitting *information*: in both the ordinary and the information-theoretic senses of the term. (As is brought out by the tacit allusion to Shannon's interpretation of a 'bit of information' as a yes-no answer to equi-probable events; see Section 1.3). This in turn further enhances the widespread conviction that

> At the moment we are trying to understand encoding and decoding of information (in the brain), mechanisms of cellular interaction, what fraction of experience is handled by a single cell, how conscious percepts result from available bits of information, and a host of other questions.[43]

Before we become embroiled in these nebulous issues, however, it would be best to close the present section by recapitulating the theme that the philosopher's greatest challenge in all this is to explain why our established concepts of mind, rule-following, and inference are not just irreducible or unalterable, but why they are irreplaceable. As far as the cognitive scientist is concerned, it seems obvious that you are going to fixate on category mistakes if you insist on dragging antiquated — philosophical — preconceptions into the debate. Science, it is believed, is searching for an alternative to our 'vernacular' — unsystematic — notions of 'mental behaviour': either by creating some new category for mentalistic/ normative phenomena, redefining mind, or if absolutely necessary, abandoning 'mind' altogether. However you approach this problem, the key is to dispense with 'anthropomorphic chauvinism'; to grasp that 'Clearly there is not just one sort of Mind.'[44] This curious thesis is intended as a self-evident interpretation of the claim that it is possible to distinguish and/or construct different kinds — e.g. mechanical versus biological — of 'information processing' systems. Thus it is that the computationalist interpretation of 'information processing' binds together the cognitivist confederacy: cognitive science embraces such different fields as AI, psychology and neurophysiology — and significantly, biology — because, in David Marr's words, AI itself is merely 'the study of complex information-processing problems that often have their roots in some aspect of biological information processing. The goal of the subject is to identify interesting and soluble information-processing problems, and solve them.'[45]

Stated in these terms the issue appears uncontentious enough, given the pivotal assumption that there could be no serious objection to the claim that information can be stored, not just in computers as much as books, but even more importantly, that 'engrams must be based on circuitry transmitting information in an organized manner.'[46] Trouble almost immediately erupts, however, with the computationalist gloss which is wedded to the notion of 'neural processing': the idea that, *qua* 'information processor', the brain/computer must be capable of framing hypotheses, deciding between competing theories, inferring from past experiences, etc. For from these purportedly bland premises Marr is soon asking us to embrace the 'categorial leap' which his theory that 'vision is the computation of a description' demands.[47] But then, were not the grounds for this 'leap' already laid down in the seemingly innocuous premiss that information can be stored in the brain in *exactly the same way* as applies to computers? If so, then our task here is to remove the logico-grammatical transgression inspiring this extended thesis, thereby subverting such extraordinary consequences as that, in even so seemingly a simple task as saying 'I see a dog', 'we are, in a sense, performing an exercise in logic of staggering complexity'.[48] Without question the only response to make to such a thesis is *philosophical*: viz. to clarify in all the various contexts why the description of the brain as an 'information-processing system' — whatever that might actually involve — *cannot* license the incoherent conclusion that

> neurons have knowledge. They have intelligence, for they are able to estimate the probability of outside events . . . And the brain gains its knowledge by a process analogous to the inductive reasoning of the classical scientific method. Neurons present arguments to the brain based on the specific features that they detect, arguments on which the brain constructs its hypothesis of perception.[49]

Here indeed is a paradigmatic example of the manner in which preconceptions can encumber the development of a burgeoning science with a surfeit of 'elegant irrelevancies'.[50]

1.3 ABYSSUS ABYSSUM INVOCAT

You cannot have a 'cognitive science of mind' without a well-defined terrain of problems for it to explore. The first step in standard introductions to the subject is thus to convince us of the extraordinary mysteries contained in even the most familiar of cognitive acts. How is it, for example, that the reader's mind is able to grasp the meaning of every word in the above sentences, much less combine these in such a way as to capture the author's semantic intentions?

> What happened took place too fast for you to notice it, but the evidence of cognitive science suggests that first you saw the variously shaped letters, then compared those images to your stored memories of printed letters (which you unerringly located somewhere in the galactic mass of your brain cells), thereby recognised the ones you had seen, recognised at the same time the word they stood for when assembled, understood its meaning, and fitted that meaning into the context of the sentence. And you did all this in a fraction of a second.[51]

There are any number of ways to respond to the myriad problems raised by this preliminary manifestation of the mind/brain's hidden mysteries, but the most important requirement is to clarify that and why these are *philosophical*, not *empirical* matters; and the traditional way of achieving this end is to indicate some of the epistemological demons which are hereby released. For instance, how we could never be certain on this picture that we had understood a sentence correctly: that our perception of the variously shaped letters was not illusory, the putative recognitional act was not delusory, the galactic mass of memory-storing brain cells was not malfunctioning, etc.; or, from another point of view, whether we could even be certain that 'the people I am acquainted with actually have a nervous system', and what such a doubt would entail on this picture.[52]

The problem with this type of *reductio*, however, is that the author runs the serious risk of being mistaken for a sceptic when what was really intended was to show why, for reasons of logical grammar, computational/neurophysiological hypotheses cannot be used to solve psychological conundrums. On the other hand we have already touched in the preceding sections on some of the

25

hazards attendant on the strategy of resorting to the charges of homunculus fallacy or category mistake; for this is all too likely to be met with the withering retort that the author is guilty of taking a metaphor over-literally. What perhaps tends to be ignored too often in all this, however, are the preliminary 'questions' from whence the mystery has sprung: the *fons et origo* for much that goes by the name of scientific controversy. In this case, it is the seemingly innocuous: 'Consider this word — the word "word" that you just read. What happened in your mind when you read it?'[53] But need anything have happened *in* my mind? Granted, I understood the word, but the danger here is trying to find some answer to the question of how I accomplished this 'feat' in terms other than would apply to the customary categorial level in which the concept of understanding is embedded; viz. in terms of the mechanics of 'neural information-processing'. The cognitive psychologist is perfectly well aware of the bruited category distinction between under standing and neural processes; but s/he suffers from a profound conceptual yearning to bridge this gap: to explain how these neurophysiological mysteries interact with and thereby constitute the cognitive results. Predictably, s/he can only succeed in such an endeavour by reading — either tacitly or explicitly — cognitive abilities into the reductionist level at which s/he longs to operate.

In order for philosophy to be truly effective against the powerful conceptual urges at work here, it is not enough to document the category mistakes which result from succumbing to this craving. While this may be a necessary part of the treatment it is not sufficient, in so far as it still leaves the cause of the disease intact: viz. the original 'questions' which the various fields banded together in the cognitive confederacy are striving to satisfy, and from which the conceptual demands of the 'paradigm' are gradually gratified. If philosophy is to be successful in its quest to expose the tissue of preconceptions sustaining the Mechanist and Cognitive Theses it must ultimately quell the yearning itself. The key to this endeavour is to show that neither neurophysiological processes nor cognitive abilities can be explained on two different levels: that, for example, what happened in my brain and how I understood the meaning of 'word' are categorially, not causally, divorced. Hence, brain processes cannot be used to explain the 'mechanics' of cognitive abilities and the mind cannot be reduced to an intermediary causal link between the two disparate types of phenomena. Our choice here — as dictated by the canons of logical grammar — is either to explain how we understood the word (in which case we shall have recourse to the concepts of learning), or else to explain the processes that occurred when our senses perceived the written symbol (in terms of the neural networks that were set into operation). To press further into the 'computational mechanics of understanding' is to create the very sort of appetency which can only be subdued, not sated. To argue thus, however, is merely to repeat the by now familiar ambitions of a Wittgensteinian critique, when what is needed is some consideration of both the problems and the fundamental notions which have inspired and sustained the Mechanist/Cognitive enterprise. This is by no means a straightforward task, however, when together these preconceptions form an interlocking network of mutually serving assumptions and affirmations.

Some idea of the scope of the problems which arise here can be seen in Jeff Coulter's penetrating discussion of James McConnell's engram experiments on worms and rats. McConnell found that rats could learn to perform certain tasks much more quickly when brain extracts from previously trained rats had been injected into their cortex. From this he concluded that 'memorial representations must be chemically encoded', from whence it is but a short step to/from Fodor's thesis that behaviour is 'the outcome of computation, and computation presupposes a medium in which to compute,'[54] or to/from the premiss/conclusion that

> The important point about MARGIE'S memory (the 'Memory Analysis, Response Generation and Inference on English' program developed by Schank, Goldman, Rieger and Riesbeck) is that it is highly active: it spontaneously reorganizes and reasons from the data within it, with results that can be drawn on in generating sensible replies to the input sentences.[55]

It is difficult to say which is the root, which the subsidiary postulates here; for together these theses constitute — at least as far as cognitivism is concerned — a seemingly impregnable defence against category objections. The important point, however, is that the philosophical issue we are confronted with in the former no less than the latter concerns the *clarification* as opposed to the *justification* of a theory or proof. The value of a paradigm — viewed sociologically, in terms of the goals and/or

27

methods uniting the research community — must be left for others to decide. However, where on the sociological view problems are assessed in light of the cultural influences guiding the paradigm, in philosophy all attention is focused on the implications of the initial 'questions' that are asked: on how and why these demand more than a metaphor for their 'resolution'.

The matter cannot be left at that, however, if we are to do justice to the genuine enigma which inspires McConnell's theory: for if we assume (for the sake of argument) that McConnell succeeded in mapping a specific ability onto an isolated chemical element, we must resolve whether this means that memories might indeed be stored in 'traces' which can be externally manipulated, physically removed, transferred from one location to another, etc. However, the philosophical problem which this theory raises is not whether any counter-explanation or superior hypothesis can be found for his results; it is *what it means* to conclude that these chemical elements can be regarded as 'encodings of information': i.e. 'memorial representations', or 'internal (neurophysiological) representations'. Here is an interpretation which assumes, among other things, that: (1) neurophysiological processes can be used to solve psychological problems; (2) it makes sense to speak of brain cells as 'symbols' or 'representations'; (3) information is stored in brain cells (in some way that is structurally isomorphic to that which is remembered); (4) this information consists in (electro-)chemically encoded data that has been transmitted to the central nervous system from the sense organs; (5) this encoded data must be decoded by the brain; (6) in which case brain cells must be capable of following internal 'neurophysiologically embodied' rules; (7) the brain must then execute inferential operations on this information; (8) hence the central nervous system performs various cognitive acts; (9) and psychological concepts can thus be used to solve neurophysiological mysteries. Are these the accretions of an evolving paradigm, or the penumbra of a deep-seated conceptual confusion?

Ironically, it is here where the computer analogy might indeed play a useful — albeit predominantly negative — expository role. As should by now be clear, the central theme of this chapter is that if we approach the issue from a mechanist preconception of computer memory, we shall swiftly be led into the beckoning arms of the Cognitive Thesis. However, from a position that proceeds with a clear understanding of the confusions that result when the logical barrier between the normativity of rule-following and intentionality of cognitive concepts versus the mechanism of computing concepts is transgressed — e.g. of the unintelligibility of attributing cognitive capacities to computer programs on the basis of the set of ciphers which trigger the various electronic signals intended by the programmer — we shall likewise be sensitive to the confusions which will immediately result if the brain is implicitly treated as an agent, responsible for reading, synthesising, and inferring from decoded information. We will be alert to the *contrasts* between cognitive acts and computer operations: e.g. to the enormous difference between saying that certain sensations can be chemically *retained*, not *encoded*; that *electrochemical impulses*, not *bits of information*, are *registered*, not *stored*; that these signals may *enable* but they do not *constitute* memories; and ultimately, that there is more than an empirical difference between saying that certain experiences can be *recalled* as opposed to *retrieved*! The remaining horn of the dilemma which looms here, however, is that without some idea of what McConnell really achieved, we seem doomed to be left once again bearing the stigma of paradigm-reactionaries; and as Coulter candidly admits, 'it is not at all clear . . . how best to interpret McConnell's findings'.[56]

Coulter's prudent advice is: 'Why not think of them just as part of the bio-chemical facilitators? "Memory" need hardly come into the interpretation'.[57] It is crucial to see what he is hinting at here. On one reading this may seem little more than a blatant evasion of the problem at hand; from another, a tacit concession to the thesis that 'memories' are simply superfluous, a primitive theoretical tool whose utility has now been eclipsed. What he intends, however, is a careful analysis of the logical grammar of the family of concepts grouped together under 'memory' *vis-à-vis* the distinct possibility that chemico-cortical transformations can be mapped onto an identifiable range of learning improvements *simpliciter*. Whereas computationalist preconceptions push us towards the thesis that, not just memories, but more importantly, the cognitive abilities with which these are commonly associated, might be 'chemically encoded', Coulter's point invokes a subtle inversion of Krech's advice to psychobiologists and brain chemists to 'Go constantly (and) look at memory in people. Know what you are studying, and whence came your question.'[58] It is a principle which

29

should be read in conjunction with Wittgenstein's warning: 'The difficult thing here is not, to dig down to the ground; no, it is to recognize the ground that lies before us as the ground.'[59] In this case an awareness of the logical grammar of 'memory' should forewarn these aspiring psychobiologists that what *they* are meant to be studying — viz. the structural and functional changes which occur in the synapses — has nothing whatsoever to do with their amateur psychological observations. Coulter's salutary physiological reminder recalls such biochemical phenomena as, for example, John Pappenheimer's vindication of the theory first postulated by Legendre and Piéron that a chemical substance — factor S — builds up in the brain of a tired animal, and that this induces sleep when transfused into another animal.

Even Coulter is not entirely happy with this argument, however, for as he subsequently remarks:

> If I suddenly remember yesterday's newspaper headline, and can quote it correctly, I may certainly be said to have stored it, and there is one sense in which this may be *literally* taken: there may be something *specific* about my CNS, brain biochemistry and/or brain functioning which enabled me to come up with *it, now*. And the question: *what* specific something(s)? is not a product of confusion but a way of beginning to decipher a *genuine* mystery of nature.[60]

In other words, philosophy cannot simply dismiss the neurophysiological puzzle of the mechanics of 'information retrieval': i.e. of how the central nervous system is — to avert to another popular metaphor — able to record and play back images, sounds, thoughts, tastes, even smells. His ultimate verdict is that this remains a completely different matter from the illicit shift from such neural phenomena to the categorial transgression involved in the attribution of cognitive abilities to the brain/computer. The question is, however, whether in conceding this point he has not already provided a sufficient chink in the argument to allow the computationalist thesis to force its way back into the theory. Once you allow the possibility of 'encoded information' to creep into physiological accounts, a decoding organ cannot be far behind. Perhaps the answer to this worry is simply that we need not voyage into the murky waters of encoded headlines in order to answer the question of how we remembered what we read; that the fact that there may be something specific about my brain biochemistry and/or brain functioning does not entail that I can in any way be said to have 'stored this information' in my central nervous system. Most important of all here is to see that what holds true for a 'neurophysiological explanation of the biomechanical facilitators for the exercise of certain (specific) *capacities*' applies *pari passu* to *memory* itself: that even if a physiologist could explain the neural processes which occurred while a subject experienced a specific memory, that would not constitute a logical/chemical analysis of the 'real' or 'neurophysiological' essence of that memory.

Such an argument will obviously invite the objection that it fails to explain how the 'memory trace' is 'registered, retained, and recalled'; but the answer to this common complaint is that 'Thinking in terms of physiological processes is extremely dangerous in connection with the clarification of conceptual problems in psychology. Thinking in physiological hypotheses deludes us sometimes with false difficulties, sometimes with false solutions.'[61] In this context we must avoid the almost pathological urge to infiltrate cognitive concepts into the neurophysiological explanation of 'memory formation' which exists, partly because of the *empirical* gap in our knowledge of the final phase in this process — viz. the mechanics of protein synthesis — and partly as a result of the misguided belief that there is a *conceptual* gap in our understanding of how information is reduced to and stored in its neural format. However, whereas the former constitutes a strictly internal scientific affair, the latter remains a subject for philosophical therapy. Thus the neurophysiologist must be especially careful, in the analysis of how the various neurons involved in learning or memory are physically and biochemically altered, that s/he does not respond or resort to psychological problems or concepts. His or her area of competence is strictly confined to the 'biomechanical facilitators' which enable us to retain our short- and long-term memories; e.g. to the electro-chemical activity which causes certain sensations or experiences. Likewise, the psychologist can have no recourse to neural nets in order to explain, for example, the results of 'reaction-time studies'. (cf. Section 2.4). Neurophysiological and psychological memory experiments proceed on categorially different planes and yield categorially divorced results. D.E.

31

Broadbent complains that 'there seems to be no sign as yet of a link between physiology and psychology of this field',[62] but this is not a consequence of their shared juvenility; rather, it is a manifestation of their conceptual autonomy; for if either discipline crosses over the logico-grammatical barrier which separates their subjects, the result will be a theoretical lacuna which *demands* a quasi-cognitive brain, busily decoding, deciding, and inferring.

Unfortunately, it is precisely here where AI has hitherto been most influential; for the Mechanist Thesis has only served to exacerbate these confusions — in the process further enhancing its own position — by applying and thence corroborating what is perhaps the basic notion underpinning this cognitive confederacy. One should, of course, always be wary of isolating a single catalyst in anything as complex as the synthesis of preconceptions and perceptions under consideration here, but it is clear that, if there is a fundamental confusion at work, it lies in the use that has been made of the concept of *information* as first developed by Shannon in the theory of communications, almost immediately (and independently) picked up by Wiener for his theory of cybernetics, and then co-opted by cognitive psychologists and neurophysiologists who discerned here the basis for a full-scale computationalist theory of the mind as embodied in the brain. A chain reaction was set off leading from Shannon's discovery that, given the proper coding techniques, messages could be separated from noise in any type of communications system, to Wiener's exploration of the similarity between noise and Brownian motion (which further strengthened the idea, implicit in Shannon's earliest thought on the subject, that these statistical methods for encoding and decoding messages could be applied to all cybernetic systems in the widest sense of the term); and finally, ending up in the belief that we can distinguish between different *kinds* of information-processing system, and hence that biologists, geneticists, neurophysiologists, psychologists, psycholinguists, and AI scientists are all engaged in the construction of theories which share the same fundamental conceptual base, and accordingly, whose individual results will have ramifying effects on the other disciplines.

The very fact that Shannon's technical definition of 'information' was isomorphic to nineteenth-century entropy equations seemed to provide compelling evidence that there was a powerful analogy between energy and information, where a high state of entropy was to be equated with noise in a communications channel and thermodynamic systems in a low state of entropy would be comparable with — if not identical to — the transmission of a message.[63] Any system or — itself a significant 'metaphorical' factor in the evolution of the paradigm — any organism which displays a tendency to become entropic must bear a fundamental analogy to a communications system. The seemingly internal link between cybernetics and information theory was obviously crucial, therefore, to the extraordinary impact of Wiener and Shannon's theory. However, if they dove-tailed so neatly, it was because they shared *ab initio* the same presuppositional base; for cybernetics assumed that there was no *a priori* obstacle to a *unified theory of self-regulating behaviour,* while information theory assumed that there was no *a priori* obstacle to a *unified theory of information processing.* The former element provided for the generalisation of the model; the latter for its conceptual substance. Together they provided a ready-made framework for struggling sciences in search of a focus. The key factor in their combined success, however, lay in their subtle intrusion of an anthropomorphic shift: the attribution of the scientist's explanatory theory to the system, subject, or organism under observation. Examples of the category transgressions which resulted are legion; but there is little point in cataloguing these when it is clear that scientists deliberately pursued such models *on the basis of* Wiener and Shannon's original theory. Thus it is with the latter that philosophy must engage if the longing for such theories is to be curtailed.

Some idea of the importance here of the philosophical 'Why?' can be gleaned from Norman Malcolm's illuminating discussion of the confusions which result when the logico-grammatical boundary between the communications-theoretic and 'ordinary' — what Shannon called the 'semantic' — notion of information is transgressed.[64] Malcolm effectively demonstrates some of the more notorious of the consequences of this infraction, such as that 'the nervous system is said to contain language, vocabulary, messages, and statements!';[65] but this resort to an exclamation mark will appear as something of a philosophical indulgence when so many scientists will treat this, not just as a perfectly intelligible hypothesis, but indeed, as a positive breakthrough. Moreover, we are given no explanation as to why this theory should have been advanced in the first

place: and commanded such immediate and widespread acceptance. One is tacitly left with the impression that here was simply
a case where scientists stumbled into this category confusion because they had misconstrued the technical nature of
Shannon's concept, largely because they had imposed incoherent cognitive demands on Shannon's information-theoretic
notion in order to satisfy their own misguided computationalist
pictures. However, such an argument is unsatisfactory as a
reading of the motives driving both Shannon and the cognitivist,
for the whole thrust of Wiener and Shannon's theory, as they
themselves conceived it, was to embrace this computational
model, both as an application and as an elucidation of his
original communications theory. Indeed, in 1943 Shannon was
already discussing with Turing their mutual conviction that
'there was nothing sacred about the brain, and that if a machine
could do as well as a brain, then it *would* be thinking'[66] — a
point of more than passing historical interest when we come to
consider the loaded terms in which he presented his revolutionary ideas in information theory.

Granted, Weaver had warned that

> '*Information*' in this theory is used in a special sense that
> must not be confused with meaning. In fact, two mesages,
> one of which is heavily loaded with meaning, and the other of
> which is pure nonsense, can be exactly equivalent from the
> present viewpoint as regards information.[67]

However, the impression that Shannon was solely occupied in
the construction of a technical concept — first advanced by
Shannon himself and then willingly accepted by scientists and
philosophers alike — is itself the source of dangerous misappre-
hensions, for Shannon set out in the *interpretation* of his theory
to do for 'information' what, in a significantly similar development, Turing had earlier undertaken for 'computation'. On the
orthodox reading, Turing presented 'a cogent and complete
logical analysis of the notion of "computation".'[68] However,
Turing is credited, not with the creation of a technical notion
(viz. 'machine computability'), but rather, with the clarification
of the real meaning of *computation*. 'Thus it was that although
people have been computing for centuries, it has only been
since 1936 that we have possessed a satisfactory answer to the
question: "What is a computation?"'[69] Similarly, 'What we ask

of information theory is that it tell us, exactly, what this thing is.'[70] To be sure, 'We may all think that we know what is meant by "information," but it is really a very subtle concept, not easy to analyse.'[71] Shannon, however, was able 'to give it a more precise meaning that will allow us to say that all life depends on a flow of information.'[72] For 'By treating information in clearly defined but wholly abstract terms, Shannon was able to generalize it, establishing laws that hold good not for a few types of information, but for all kinds, everywhere.'[73] All of the terms in his theory convey Shannon's conviction that such was indeed the prospect opened up by his definition of 'information': a theme which is broadcast in the very term 'information theory'. Certainly, philosophy is not empowered to forbid such a technical definition to go by the name of 'information': but it is fully within its compass to advertise the dangers of using pre-existing semantic names for the statistical network of concepts contained in Shannon's theory. Moreover, any philosophical anxieties on this score have been more than amply justified.

From the premiss that messages could be electronically coded into strings of binary digits, Shannon promptly shifted to the conclusion that *all* information must be so encoded; but a coded message cannot be treated as the sum total of its binary digits: the message is the information that has been thus encoded. (A telegraphist who passed on messages to his customers in the Morse code in which they had been transmitted would not stay a telegraphist for long.) Whether a message composed of 100 binary digits contains less information than one with 200 depends entirely upon the intentions of the sender, not on the vagaries of his or her code; for a binary digit is the smallest unit of the digital *code*, not of the information transmitted by the code. The fact that a message can be coded into n binary digits clearly no more entails that the message contains n 'units of information' than the fact that a sentence is composed of n letters entails that a letter is the minimal unit of the information conveyed by the proposition. Not surprisingly, however, communications theorists frequently conflate these two ideas — despite their emphasis on the importance of the distinction between binary digits and 'bits of information'; indeed, the latter becomes an illustration of what is meant by the former. For the prose of the theory misguidedly reads the amount of information yielded by the theory's statistical rules (e.g. that some letters are likely to appear more often than others) into the binary digits/letters themselves, thereby resulting in the conclusion that 'the amount of information in a letter, word, or sentence is inversely proportional to the frequency with which it occurs.'[74] However, the reason why there is no suggestion here that we are dealing with a rarified meaning of 'information' is simply that, from the start, Shannon's interpretation of his theory assumed that it makes sense to speak of the 'amount of information' contained in any message; that information is something which can be quantified and broken down into constituent elements.

Fred I. Dretske's subtle defence against the (by now familiar) complaint that 'information theory' is a misnomer, in so far as the theory is really one about 'signal transmission', brings out — and at the same time exemplifies the problems involved in — the semantic undertones suffusing information theory. For his is a defence which assumes from the outset precisely that which it purports to establish. Dretske explains that 'Communication theory does not tell us what information is. It ignores questions having to do with the *content* of signals, what *specific information* they carry, in order to describe *how much* information they carry.'[75] However, in arguing thus he has *already* assumed the categorial shift from the ordinary to the technical notion of 'information' without recognising let alone acknowledging the fact: a manoeuvre which is crucial to the subsequent elaboration of a 'semantic theory of information'. He concedes that 'there is certainly a distinction between the message a signal bears and the signal bearing that message.' Yet 'A distinction can also be drawn between the water a bucket holds and the bucket that holds the water'; and if we 'can measure the bucket to find out how much water it contains', they why 'can one not determine a signal's informational content, or at least how much information it contains, by taking the measure of the signal itself?'[76] However, if we draw a parallel between information and water in ordinary terms, this can in no way license the conclusion that, on a par with ounces, there are *units* of information. What the argument really shows is that there is an analogy between 'message' as information-theoretically conceived (viz. 'a function of the actual possibilities that exist at *s* and the conditional probabilities of these various possibilities after the message has been received'[77]) and the water in the bucket. Only in this case the analogy masks over a special sort of problem, for the argument is intended to substantiate the claim that information *per se* can be quantified; but the analogy proceeds on the prior understanding that probability measures can be quantified, as of course, is *per definiens* the case. Thus the argument assumes on the strength of the analogy — and not the reverse — that all messages can be quantified in binary units (although the use of the latter would be but one of many possible conventions, *pace* the use of the imperial system). The analogy only operates, therefore, at a level where 'messages' can be described in communications-theoretic terms (where the 'amount of information' contained in a 'message' is a function of the number of binary decisions necessary to reduce *n* possibilities to 1, together with the various constraints on the communications channel, the nature of the code adopted, background knowledge, etc.).

Dretske prefaces his account with the claim that 'what information a signal carries is what we can learn from it'.[78] Disregarding the precipitate intrusion of 'signals' here, the information theorist's basic problem is that what the two communities of speakers involved — viz. language users versus communications engineers — can learn from a given proposition is not *qualitatively*, it is *categorially* different (just as we saw in the preceding section that the same utterance can be described on categorially disparate levels). There are really two separate 'messages' involved: that which is understood by language-speakers, and the information theorist's interpretation. Dretske maintains, for example, that 'if I tell you that Denny lives on Adams Street in Madison, Wisconsin, I give you *more information* than if I tell you, simply, that he lives in Madison, Wisconsin.' That is certainly the case, but it in no way licenses the shift to *quantifying* that information — for the added information you have given *me* is the name of the street on which Denny lives, not a factor of his possible dwelling-places. In order to create the impression that such an example illustrates the quantifiable essence of information, the argument implicitly moves from this information to the 'information$_{IT}$' that 'the number of possible places to live in Madison, Wisconsin, is greater than the number of possible places to live on Adams Street in Madison, Wisconsin.'[79] In other words, the argument switches to the level of the information theorist's interpretation, and the only way to merge these disparate categories without describing 'information$_{IT}$' in terms which no language-speaker could understand is by transforming

37

language-speakers into *tacit* communications engineers: i.e. into 'information-processing systems'.[80]

When the information theorist discourses on, for example, 'the amount of information' contained in a page of English prose, s/he may ostensibly be concerned with the probability that a given letter will appear — or more generally, the probability that a sequence of signals will appear as measured by the previous distribution of signals — but the semantic terms in which the theory is couched ensures that the information theorist is no longer confined to jejune communications problems. For 'message' as now conceived has become an amorphous concept embracing 'any kind of information system in which (signals) are sent from a source to a receiver'.[81] 'Communication' itself is no longer 'confined to radios, telephones, and television channels: it occurs in nature, wherever life exists. The genes are a system for sending chemical messages to the protein factories of the cell, instructing them to make a living organism. The human being is the most complex communications network on Earth, and language is a code which preserves the orderly structure of the messages of speech in ways so ingenious that they are still not fully understood.'[82] Nature as well as man must therefore have evolved complex redundancy codes to separate out encoded binary signals from noise. In conveying thoughts, language-speakers — *qua* 'information source' — are literally transmitting messages, and so in understanding thoughts they must — *qua* 'receiver' — be decoding those messages. The fact that they are unconscious of the latter process is hardly an obstacle to the theory; rather, it merely confirms the neurophysiological levels at which such operations must occur. So it is that the emergence of an 'information-processing' brain which 'contains language, vocabulary, messages, and statements' is no idle categorial slip: the very act of presenting information in 'information-theoretic' terms eventually demands just such a framework.

The theory still does not possess the vital ingredient to lift it off the ground, however; for at the most we are now left with a motley of communications engineering homunculi, all of them expert in the finer details of information theory. In order to complete the attribution of these information-processing operations to the systems and organisms themselves, a species of 'rule' was needed, suitable for any cybernetic mechanism, organism, or indeed, organ. Fortunately, this lay ready to hand in Turing's 'logical analysis' of computation, for Turing had purportedly reduced algorithms to complex systems of meaningless sub-rules, each of which can be applied purely mechanically. Moreover, if 'what Turing did was to analyze the human calculating act and arrive at a number of simple operations which are obviously mechanical in nature and yet can be shown to be capable of being combined to perform arbitrarily complex mechanical operations,'[83] then any information-processing system could be said to be 'following rules' in some version of this 'non-normative' sense: an assumption which is both the source and the prime significance of the Mechanist Thesis. What this means is that wherever the operations of a system or organism can be *mapped* onto an algorithm, there is no logical obstacle to the transfer of that algorithm to the system or organism itself. Wittgenstein's warning that the possibility of mapping Zulu war dances on to chess in no way shows that the warriors are (tacit) chess-players holds no water here. Rather, the issue is simply whether computers could possibly be said to be capable of a feat that is to be denied to the brains that had created them? In strikingly Platonist terms, the scientist emerges not as *creator* but rather as *discoverer* of these programs in the system under observation.

In the case of DNA, for example, 'The algorithm would be a kind of program, instructing certain combinations of genes to turn on or turn off at specific times, and it would be stored in the DNA text as information'.[84] Admittedly,

> Most biologists (may) tend to see themselves as describers of nature, collectors of facts which will lead to a good experiment. Yet if DNA is an information process, and a theory of information exists, then it is reasonable to suppose that scientists can at least make a start of sketching out a theory of living organisms.[85]

In other words, the biologist is not engaged in the construction of statistical rules for measuring the frequency of AGTC in a DNA chain; rather, s/he is *discovering the algorithms* whereby biological information-processing systems regulate their (evolutionary) 'behaviour'. Indeed, on this picture, 'The discovery that [one of the smallest bacterial viruses, ØX174] stores information by means of a DNA text so cunningly composed as to tax the ingenuity of a master anagrammatist came as a revelation.'[86] Now, perhaps, would be the appropriate moment for exclamation marks, were it not for the complex role which AI has played as both benefactor and beneficiary in the evolution of this computationalist 'paradigm'. Indeed, Ayer's counsel that we do not concentrate unduly on the Mechanist Thesis begins to take on ever greater and more ominous significance; for if nothing else, it should serve as a warning that we do not underestimate the strength of the filaments binding AI to the cognitive confederacy.

1.4 THE FATHERS OF CONFEDERATION

We began this chapter by querying whether philosophers must bow to the scientistic judgement inspired by Exodus 20:5, but have now reached the point where we can consider whether it is not cognitive scientists who should take heed of the proverb. Here is a topic where the sociology of science can indeed perform an inestimable service for our understanding of the complex interplay of factors which fathered not simply the mechanist framework for AI, but on that basis, bequeathed the computationalist legacy which has inspired their cognitive sons — verily, unto the Fifth Generation and beyond. A promising start to this undertaking can be found in Chapter 10 of Steve J. Heims' *John von Neumann and Norbert Wiener*.[87] The full chronicle of the genesis of cognitive science will only be complete when its origins have been traced back to the revolution in axiomatics which occurred at the turn of the century. However, Heims makes a valuable contribution with his account of how, some 40 years later, the unique demands of the war effort had created a climate of interdisciplinary familiarity and co-operation, not just between scientists from disparate backgrounds, but equally importantly, between science in general and government (funding!) agencies. This renewed appetite for scientific cross-fertilisation found its natural gratification in AI: an area where enlightened confrères could pursue in more abstract terms the problems which had been opened up by their work during the war in communications theory and logistics. It was at the 'Macy Foundation' conferences beginning in 1946 — originally called the 'Conference for Circular Causal and Feedback Mechanisms in Biological and Social Systems' and subsequently shortened to 'Conference on Cybernetics' — that the founding fathers mapped out the unwritten constitution for their cognitive confederacy.

Given the appointed purpose of these colloquia, there was an inescapable pull for all concerned to explore the ways in which they could adapt and employ each other's theories; and given the predominant role which Norbert Wiener played in the proceedings, it was inevitable that cybernetic concepts should come to provide the framework for this exercise. However, that is hardly the entire story, for the controlling concept was that of 'information', not 'servo-mechanisms' (and one might well ponder whether there would be less confusion today had it been the latter which prevailed — or better still, no single concept at all). The list of scientists who attended these gatherings reads as an honours roll of the leading architects of cognitive studies.[88] However, even though the stature of these figures precludes any simplistic reading of the guiding influences of AI, for the purposes of this chapter it is two non-participants, Turing and Shannon, who stand out in sharp relief against this imposing cybernetic background. Where each of the star performers at the Macy conferences supplied a different element for the mechanist doctrine which evolved from their collusion — e.g. von Neumann, the idea of formal logical models of the brain; Wiener, the general notion of the 'feedback' mechanisms underpinning 'purposive behaviour'; McCulloch and Pitts, the axiomatic theory of formal neural networks; and Bateson, the biological and anthropological extension of the model — it was Turing and Shannon who provided the key ingredients which enabled the admixture to gel. However, for partly sociological and partly philosophical reasons, the full significance of this influence was only partially understood, much less noted by the Macy confederates.

To be sure, in 'The general and logical theory of automata', von Neumann paid tribute to Turing's pioneering analysis of formal systems which can simulate any 'mechanically effective computation', but he then passed on to what he saw as the immediate precursor for much of what he aspired towards in the theory of automata: McCulloch and Pitts's 'A Logical Calculus of the Ideas Immanent in Nervous Activity'.[89] However, in the discussion which followed, McCulloch revealed that their own paper had been inspired by Turing's 'On Computable Numbers'.[90] As for Shannon, it is generally agreed that it was his work in communications theory more than Wiener's which shaped the course of the information-theoretic notions which provided the raw materials on which to run Turing's 'machines'. However, apart from the unnecessary contention over priority that such claims invariably provoke, what really matters here is the fact that Turing and Shannon's collaboration — formally cemented during Turing's visit to the Bell Laboratories in 1943 — was based on the most fundamental of the many ideas that they exchanged: the notion of a *bit*. It is from this seemingly insignificant origin that we can begin to see why AI has served as very much the binding agent in the volatile cognitive diffusion outlined in the preceding sections: to a considerable extent a direct result of Turing and Shannon's mutual interest in the role that computers could play in our understanding of the human brain.[91] For they shared and shaped the assumption that computer programs would enable scientists to 'mimic the effects of rules built into the structure of matter.'[92] They based this conviction on the analogy which each drew between a nerve impulse and a binary digit: if the latter could be defined as the carrier of 'units of information', then so also could the former, thus enabling science to bypass 'semantic' quibbles over the meaning of 'thought' and henceforward classify any structure exhibiting cybernetic symptoms as a (sub-)species of 'information-processing' system.

Turing was clearly the more extreme of the two, and perhaps for that reason, the more influential at the outset. In 'Can a Machine Think?' Turing boldly addressed the standard humanist objections to the Mechanist Thesis with all of the gusto which only a 'hard-headed thinker' can muster; in 'A Chess-playing machine', Shannon would only commit himself to the point that 'several large-scale electronic computing machines have been constructed which are capable of something very close to the reasoning process.'[93] However, while circumspect about the philosophical minefield surrounding the concept of thought, Shannon saw no objection to referring to the rules which computers can follow, and thus, to the notion of 'learning' and 'chess-playing' systems.[94] Our concern here, however, is not so much with the positions which Turing and Shannon took on the Mechanist Thesis, as the far-reaching influence which they had on the emergence of computationalism. Here also, Turing was the more extreme of the two. As far as he was concerned, the logical structure of the brain could be described in terms of 'Turing machines' simply because each represented discrete logical systems.[95] Shannon, however, was not interested in a crude mechanist interpretation of mental/neural events. On the contrary, he accepted — or perhaps inspired — the now prevailing belief that we can 'learn more about the human brain by studying the ways in which it differs from the computer than by looking for resemblances between the two.'[96] On the surface this may seem a harmless enough proposal, whose worst fault would turn out to be its fruitlessness. However, together with the 'mechanical rules' with which Turing proposed to destroy the (human) mind's cognitive hegemony, the argument introduced a subtle and insidious assumption into the Macy forum.

Shannon's point was that the various 'information-processing systems' — in the generalised sense in which he interpreted the concept — will operate on vastly different 'internal rules'; hence, they will all display their own specific 'functional organisation', and by comparing these to computer programs we can hope to highlight the unique characteristics of each. However, the mere fact of contrasting the brain with a computer program entails that the same conceptual scaffolding applies to both: viz, that the brain 'is a system for coding and organizing information, and can best be understood with that framework of ideas.'[97] With this in place as the starting-point the path to, for example, Richard Gregory's inferential theory of perception is remarkably straightforward, as are also the host of paradoxes which this entails.[98] What is seldom noticed in this debate, however, is that even those most enthusiastic about the scope of neurophysiological applications for the computer 'metaphor' remain circumspect about its literal incorporation; while even those who are most dubious about its utility remain committed to Shannon's information-theoretic approach to cognitive abilities. Thus, Young warns that 'The analogy of computer programs can indeed be helpful — but only in the most general sense'.[99] While despite his preference for 'the assumptions to stem from the results rather than the conclusions from the assumptions',[100] Weiskrantz remains committed to the premiss that, for example, 'patients with bitemporal lesions appear to find it difficult if not impossible to put new information into storage'.[101] Much as he may dislike the computer analogy, Weiskrantz has already accepted with this premiss the seeds for Young's 'decoding' — and thence, computing — brain: for if a computer can be said to be following rules, then why not the brain?

For those not fully conversant with the pernicious influence which the Mechanist Metaphor can — and to some extent already has — exercised on scientific methodologies the first part of Weiskrantz's paper should be required reading; but it is no less important to reverse the picture and consider the dangers which this conceptual incest poses to the development of AI itself. Campbell describes how

> In artificial intelligence research, scientists are learning some lessons from the fact that the brain processes information in ways which are peculiarly, even perversely human, rather than mechanical in the old sense. For example, they have made the paradoxical discovery that forgetting serves a very important function, and is a by-product of learning. Computers are now being programmed to forget selectively, as the brain does, rather than store every item of information in its memory.[102]

However, if neither the brain nor a computer 'forgets', where exactly is the analogy leading both? As far as AI is concerned, it is to a series of reverses extending from the implementation of expert systems to the construction of mechanical perceptual systems.[103] While as far as neurophysiology is concerned, the answer is: to *mystery* — and thus a reminder of the limits of reason. It is fascinating to see how Granit prefaces his account of 'the purposive brain' (and returns throughout his exposition of the theory) with Berzelius's admonition that 'We cannot reach farther than to understand what can be understood and realise what we cannot understand.'[104] In Wittgensteinian terms this answer leads to the most damning of any charge that can be found in the philosopher's lexicon; for it results in a metaphysics of the mind (in the literal sense of the term). However, thanks to the services proffered by AI scientists, even this charge can be imprudently dismissed on the 'vanguard' model of paradigm evolution, which has boldly dispensed with both sides of this 'antiquated' allegation.

Thus the AI scientist advises that 'The only alternative' to classical mentalist or behaviouralist accounts of the nature of mind

until recently has appeared to be to locate mind in brain

matter — but this ignores important category distinctions: although neuronal states, events or processes may correlate with my being conscious, they are not themselves conscious-ness. . . . Yet any other attempt to identify a referent for 'mind', 'conscious', 'pain' etc. has, until recently, looked like an attempt to populate the world with mysterious, inaccessible metaphysically unjustified entities.

However, 'what is different now is that computing science has provided us with the concept of a *virtual machine*'. This is not simply a computer program, however; rather, the latter is itself but one example from a 'richly structured space' ranging from 'simple servo-mechanisms like thermostats' to 'the simplest organisms'. In general terms a 'virtual machine' is essentially any 'behaving system' which follows an internal set of mechanical rules for obtaining a specific output from a specific input. Hence, 'a virtual machine has much in common with the kind of formal system studied by mathematicians or logicians'. However, 'Work in Artificial Intelligence has shown that some virtual machines can produce behaviour which previously had been associated only with minds of living things, such as producing or understanding language, solving problems, making and executing plans, learning new strategies, playing games.'[105] The idea is yet another version of the Turing machine, only this time one that is explicitly constructed along information-theoretic lines; for the algorithms which these 'virtual machines' follow are no longer confined to the barren precincts of predicate logic which seemed the fate of Turing's 'slaves'. Placed within the context of Shannon's theory, a vast range of cognitive processes lies waiting, opened up by the emergence of complex systems of meaningless sub-rules for the transmission and processing of information: the joint product of Turing and Shannon's invention.

The trouble with concepts such as 'Turing/virtual machines' for both the sociologist and the philosopher of science, however, is that where the cognitivist treats this as the irrefragable foundation for his theory, the philosopher sees in it the starting-point for his *reductio*. Needless to say, there is bound to be a serious communication problem when two parties see in one and the same concept the completely antithetical conclusion. In the philosophy of mathematics, for example, we frequently seek to clarify the conceptual confusions of a particular theory by first depicting the metaphysical consequences which it entails and then working our way backwards to the sources of the theory. For all too often, the most difficult element of the philosophy of mathematics is to convince mathematicians that a crime has indeed been committed. Yet to accomplish this we blithely regard platonism as the *corpus delicti* of our investigations and the mathematician found indulging in transcendental deliberations as caught *in flagrante delicto*. Naturally a platonist will find this procedure preposterous; after all, we are denying him the guiding principle of his *Weltanschauung*. However, what is far more worrying is that the grand metaphysical 'discoveries' have all too often been founded on the most substantial of mathematical or logical theorems. Constructivists may wish to dissociate themselves from the platonist consequences which, for example, Douglas Hofstadter or Michael Guillen draw form Gödel's second incompleteness theorem, but the issue they must ultimately address is whether Gödel's proof does indeed constitute a validation of platonism (as no less an authority than Gödel himself believed), or whether these metaphysical theses themselves constitute grounds for reconsidering the standard meta-mathematical interpretation of Gödel's second incompleteness theorem (as Gödel conceived it).[106]

It is bad enough when two sides are diametrically opposed on their interpretation of the most basic notion in their dispute, but to make matters worse, the same problem arises at virtually every step in the philosopher's downward investigation of the mechanist confusions which have hitherto suborned AI. For the whole process must be repeated at, for example, the level of information theory; here too the argument sets out from the very premiss on which the philosopher seizes as the crux of his or her *reductio*. Moreover, to complicate the situation still further, the philosopher must also contend with the theory's upward dynamic, in as much as 'virtual machines' can be invoked to corroborate the assumptions of information theory. What is most troubling in this issue is thus the manner in which the Mechanist Thesis functions as both the product and confirmation of the Cognitive Thesis that *thought* can be analysed as a series of sub-routines. The shibboleth which joins this disparate company is 'cognition' itself: a euphemism in the computationalist's hands for the shift from thinking to 'information processing'. And the starting-point for this grand design is the idea that, *qua* 'general information-processing machine [a computer] can perform operations analogous to human thinking'. Not surprisingly,

> By the 1950s the inevitable happened: psychologists began to turn the analogy around and liken the mind to an intelligent machine — that is, view it as an extremely sophisticated information-processing mechanism, and all the devices inside it as stages in an information-processing sequence.[107]

Here is a sterling example of the manner in which a concept can be drawn *from* AI, and yet it is the full range of cognitive abilities associated with thought which licenses the application of the AI term in this cognitive context. Hence, we have a case where the ties between AI and cognitive science melt into one another: a consequence of the fact that, as we saw in the preceding section, as soon as you begin to speak of 'information processing' you have opened Pandora's box.

On the 'information-processing' model of the brain we are quickly brought up against a seemingly impassable barrier; for if we assume,

> as seems inevitable, that the brain encodes and processes information in some electrical-chemical code, which we will call 'neuronese' the question arises as to how and why it is that I understand English but haven't a clue as to the semantic and syntactic features of neuronese.[108]

Stated in these terms one might well be forgiven for supposing that the author intends this as the premiss of a *reductio*. However, thanks to AI such is not at all the case; for 'Without answering this question directly one can see how computers give a purchase on the possibility of a solution to this conundrum.' In other words,

> Both humans and sophisticated computers can receive inputs and represent outputs in a language utterly different from the languages in which they process these inputs and outputs. This suggests that human systems, like computational systems, require interpreters and involve several different levels of processing.

The computer may not serve as an adequate explanation for how the brain actually functions, but it at least provides a perspicuous example of the type of hierarchical structure involved in information-processing systems. Hence, 'the mysterious fact that I understand English while my brain understands neuronese is explained by the fact that an entire system (in this case me) and a subsystem (my brain) can have different properties.'[109] However, what licenses the various cognitive terms operating in this picture of computer systems? Here we have shifted from the point that we can use computers to process information (converting high-level languages into machine code by means of symbolic assemblers and compilers) to the mechanist assumption that it is the system itself which is *following rules* and *executing instructions*; which *interprets* and *infers from* the information we have encoded.[110] Further, this transgression of the logico-grammatical divide between normative and mechanical concepts is itself based on the premiss that the brain operates as an information-processing system, and since the latter *must* enable us to understand English, then a similar process applies to computer systems.

Thus it is that AI operates as both rationale and forum for computationalist preconceptions, for AI provides the perfect vehicle for cognitive scientists to test their transcendental deductions. This is obviously most evident in one of the mainsprings of cognitive psychology: reaction-time studies. Here

> little experiments are performed on the basis of which psychologists draw transcendental inferences to the effect that people perform exhaustive serial searches on lists of digits, remember words better on the basis of first letters than third letters, cannot hear speech as meaningless, and so on.[111]

The crucial premiss underlying these studies is that, if you can map mathematical functions onto a sequence of reaction times, then that function must actually be a manifestation of the computational operations performed by the mind/ brain. In a classic experiment R.N. Shepard and J. Metzler presented subjects with three pairs of two-dimensional pictures of rotated three-dimensional objects and then asked whether the paired objects were the same. They 'found that the reaction times were a linear function of the degree of rotation', and they based this conclusion on the

> Kantian question: How are these data possible? Shepard and Metzler's transcendental inference is that they are possible only if people in fact mentally represent and then rotate the figures in some medium which is representationally analogous to three-dimensional space.[112]

Such experiments are encumbered from the start with a grave methodological problem; for 'Transcendental reasoning is always radically underdetermined by the available evidence. There is always an enormous number, in some cases an infinite number, of possible hypotheses compatible with some set of reaction-time data.'[113] Once again, however, AI comes to the rescue, this time because

> it promotes the building of more comprehensive models that, on the one hand, are required to come to grips with the solid but piecemeal experimental data, and that, on the other hand, will foster new explanations, predictions, and questions about how the mind works.

These 'predictions can then be tested in artificial or natural settings — that is, on computers or people.'[114]

The process thus set off exemplifies the inextricability of the assumptions which underpin AI and cognitive psychology. To begin with, 'AI requires that the psychologist design formal models of alleged cognitive processes and run them on computers.' But then, 'Actually running such programs on computers has the further advantage of making sure that the alleged processes are, in fact, computationally realisable.'[115] The comparison presupposes, therefore, the mutual intelligibility of these interlocking assumptions: the Mechanist Thesis certifies that the minds embodied in brains are information-processing systems, the results of which substantiate the prior supposition that computers are information-processing systems. (For example, the computationalist picture of pattern recognition can be used to launch the Mechanical Perception Thesis, which in turn provides working models for the possible computational operations performed in the mind/brain.[116]) This objection is not intended to impugn the possible significance of reaction-time studies, however: only the transcendental deduction which imposes a computationalist model on the data generated (although without this framework it becomes doubtful whether reaction-time studies would retain the same significance for cognitive psychologists). The fact that we may be able to discern certain empirical regularities (e.g. temporal) in reaction-time studies in no way licenses or demands the *categorial* shift to a computationalist explanation. If there is a genuine puzzle in the results generated by such experiments, that is something which must be examined and explained *on the same level* as that on which the experiment was conducted. All too often, however, that which baffles is itself merely a consequence of computationalist preconceptions.

For example, perhaps the most interesting aspect of Shepard and Metzler's experiment is the fact that 'Almost all subjects claimed that they performed the matches or discovered the mismatch by mentally rotating one of the figures until it was congruent with the other.'[117] Stated in these terms, this obviously poses no embarrassment for the computationalist theory; for 'transcendental deduction' is really just another name for forcing the facts to satisfy the demands of the model. We are told that 'Shepard and Metzler's interpretation that some kind of spatial transformation takes place is more plausible the more credibility we assign to the subjects' reports that they are in fact mentally rotating images.'[118] That is, mental spatial representations are to be treated as isomorphic with AI models, even though the manner in which these are manipulated and identified may be functionally different. However, the real question here is how it is that the subjects' reports that they could only solve the problem by *visualising* the rotated objects pictured was translated into the notion of *rotating mental representations of the object*? What does it mean to speak of a mental image as a 'representation', let alone of 'rotating an image' or 'matching an image with an object'? The fact that it may have taken the subjects twice as long to visualise figures at a 160-degree rotation distance as those at an 80-degree rotation distance does not mean that it took them twice as long 'mentally to rotate' images of those figures in their minds/brains! Here is a blatant example of reading the investigator's perspective into the subjects' performance, thus presenting the data in terms which implicitly pre-suppose a computationalist framework. Nor is it an envious position to be in; for the latter only succeeds in creating a new host of *mysteries*, not in providing an *explanation* for any temporal patterns recorded: e.g. where and how these 'representations' are 'stored', how they are 'rotated', whether two-dimensional 'representations' are 'rotated' in the same way as 3-D ones (and indeed, if there is an upper limit to the mind/brain's ability to 'rotate' n-dimensional 'representations'), how they are 'matched' and how this result is communicated to our consciousness, how the mind/brain can be certain that a match has been obtained, etc.

To be sure, the cognitive scientist is prepared to concede that anomalies will continue to arise on this approach. Pattern-recognition systems, for example, remain confined to 'toy domains', and at that, hopelessly slow in relation to human pattern-recognition; and neurophysiology has still to sight any of these cognitive centres, although for computationalists their tracks are ubiquitous. However, these set-backs are all to be explained *empirically*, e.g. in terms of the limitations of serial as opposed to parallel processing, the superiority of the brain's 'matching algorithms' and analogue strategies, or the complexity of neural networks. Among other outcomes this has meant that an inordinate amount of both AI and cognitive scientists' time has already been spent in constructing models to satisfy the computationalist demands of their transcendental reasonings. It was precisely this type of hazard which Weiskrantz had in mind when he begged for the assumptions to stem from the results rather than the conclusions from the assumptions. Yet the scientist often finds him or herself powerless to reverse this trend; for if s/he challenges the cogency of an analogy s/he more than anyone runs the risk of being branded as a 'paradigm reactionary'; particularly when s/he shares the conceptual background for the paradigm which so alarms him or her. Yet for the same reason s/he must be discreet in any support s/he might wish to lend to philosophical efforts to expose the sources of the 'elegant irrelevancies' besetting his or her field. In particular s/he must be circumspect about sanctioning the philosopher's *reductios*, even though these are but a method of reversing the process that has occurred in the 'abuse of metaphors' by using the results to clarify the nature of the assumptions that have inspired the 'myth'. For here the serious risk is run of questioning the canons of the scientists' science; and as everyone knows, heresy is universally regarded as a far more heinous crime than prejudice.

If there is something inherently dissatisfying in basing our *reductios* on scientists' proudest achievements, however, perhaps the issue can be reduced to terms which not even the cognitive scientist can afford to ignore; that is, to the logical obstacles which have been built into the enterprise from the outset. The issue does indeed rest on a vicious form of circularity: one where mechanist assumptions have been freely transported from AI to become working hypotheses in cognitive science, itself the vindication of the Mechanist Thesis. Yet another answer to the question, 'Where is all this leading to?' is thus: to a conception of *'mental states'* where cognitivists from whatever scientific persuasion can meet each other on common ground. Here they can at least communicate with each other, if with no one else. For the danger now is that AI models designed specifically to meet cognitive requirements might at best realise computationalist preconceptions, but little else besides. However, even this way of stating the matter runs the risk of being unduly charitable to cognitive science. In his Postscript to *Eye and Brain* Richard Gregory reports that, because 'performance does not justify the claim that Machine Perception matches even the modest biological perception given by simple brains . . . (s)ome of the biological renegade emigrés into "intellectual technology" (including the author) returned to mice and men.'[119] Even Margaret Boden has recently acknowledged the existence of a growing misgiving that AI 'has not been as helpful to working psychologists as its supporters initially hoped.'[120] Both remain convinced, however, that the obstacles hindering the progress of the cognitive sciences are purely contingent.

In pleading tones Gregory asks: 'Given that eyes and brains work, why should it be *impossible* to make artificial intelligence? Is there something intrinsically unique about brains?'[121] The foregoing paper has endeavoured to show how very misleading such a question is. The computationalist can be counted upon to respond, however: 'What do you mean, it is misleading? Where does it lead you to?' Nearly 50 years ago Wittgenstein complained that the problem with such pictures is that they command an attitude such as that epitomised by Sir James Jeans's *The Mysterious Universe*: a title which, so Wittgenstein believed, 'includes a kind of idol worship, the idol being Science and the Scientist.'[122] Technology may have advanced at a terrific pace since then, but the mysteries remains no less obdurate. In 'The Confounded Eye' Gregory begins:

Eye and brain combine to give detailed knowledge of objects beyond the range of probing touch. Just how this is achieved remains in many ways mysterious; but we know now that specific features of objects are selected and combined to give an internal account of the object world.[123]

However, the latter is not *known*: it is *postulated*; and as the former part of this statement reluctantly acknowledges, cognitive scientists are still no closer to unravelling the mysteries of this 'process' than was Helmholtz over a century ago. Granted, an embryonic science cannot allow the inevitable disappointments which haunt a new venture to daunt its resoluteness; but when theories are parasitic on each other, they must share the burden equally when conceptual confusions are the cause of their joint frustration. Moreover, philosophy's responsibility is to hasten this result by exposing whenever necessary the 'curious attitude scientists have — : "We still don't know that; but it is knowable and it is only a matter of time before we get to know it!" As if that went without saying.'[124]

NOTES AND REFERENCES

1. In his preface to J. David Bolter's *Turing's man* (Penguin, Harmondsworth, 1986), p. ix.
2. James Fleck, 'Development and establishment in Artificial Intelligence' in Norbert Elias, Herminio Martins and Richard Whitley (eds) *Scientific establishments and hierarchies, Sociology of the sciences*, vol. VI, (Reidel, Dordrecht, 1982) p. 181.
3. Thomas S. Kuhn, *The structure of scientific revolutions*, 2nd edn (University of Chicago Press, Chicago, 1970), p. 19.
4. John Haugeland, *Artificial Intelligence: the very idea* (MIT Press, Cambridge, Mass., 1985), p. 5.
5. Cf. ibid., p. 6.
6. See the penetrating dialogue between the Duhemist and Campbellian in Mary B. Hesse's *Models and analogies in science* (University of Notre Dame Press, Notre Dame, 1966).
7. Haugeland, *Artificial Intelligence: the very idea*, p. 2.
8. Ibid., pp. xii–xiii.
9. Bolter, *Turing's man*, p. 13.
10. See Kenny's seminal paper, 'The homunculus fallacy', in his *The legacy of Wittgenstein* (Basil Blackwell, Oxford, 1984).
11. Ibid., p. 14.

12. See Margaret Masterman, 'The nature of a paradigm' in Imre Lakatos and Alan Musgrave (eds) *Criticism and the growth of knowledge* (Cambridge University Press, Cambridge, 1970), pp. 61ff.

13. Cf. D.O. Edge on the 'perceptual' role of metaphor, in 'Technological metaphor' in D.O. Edge and J.N. Wolfe (eds) *Meaning and control: essays in social aspects of science and technology* (Tavistock Publications, London, 1973), pp. 35f; and Masterman on the role of paradigms as a 'way of seeing' in § 4 of her 'The nature of a paradigm'.

14. See Douglas Berggren, 'The use and abuse of metaphor I and II', *Review of Metaphysics* vol. 16 (1962–3), pp. 237–58, 450–72.

15. Kuhn, *The structure of scientific revolutions*, p. 37.

16. See S.G. Shanker, *Wittgenstein and the turning-point in the philosophy of mathematics* (Croom Helm, London, 1986), Chapter III, pp. 75–119.

17. Kuhn, *The structure of scientific revolutions*, p. 37.

18. Cf. G. Nigel Gilbert and Michael Mulkay's enlightening record of this very point in *Opening Pandora's Box: a sociological analysis of scientists' discourse* (Cambridge University Press, Cambridge, 1984).

19. See S. G. Shanker, 'The decline and fall of the Mechanist Metaphor' in Rainer Born (ed.), *The case against AI* (Croom Helm, London, 1987), pp. 72–131.

20. Although even this might conceivably be incorporated into Kuhn's amorphous conception of 'paradigm'; see Masterman's definitions (3), (8), and (21), on pp. 62–4 of 'The nature of a paradigm'.

21. Ibid., p. 52.

22. Steve Woolgar, 'Why not a sociology of machines? The case of Sociology and Artificial Intelligence', *Sociology*, vol. 19, no. 4, (1985) pp. 557–72.

23. Bearing in mind that, as is pointed out in the OED, 'natural philosophy' and 'science' were virtually synonymous in middle English.

24. Cf. Daniel C. Dennett, 'Artificial Intelligence as philosophy and as psychology', in his *Brainstorms* (Harvester Press, Brighton, 1985); John Cohen, *Human robots in myth and science* (George Allen & Unwin, London, 1966).

25. See D.A. Schon, *Invention and the evolution of ideas* (Tavistock, London, 1967).

26. Donald Michie and Rory Johnston, *The creative computer: machine intelligence and human knowledge* (Penguin, Harmondsworth, 1986), p. 18.

27. Haugeland, *Artificial Intelligence: the very idea*, p. 2.

28. R.L. Gregory, *Eye and brain; the psychology of seeing*, 3rd edn (Weidenfeld and Nicholson, London, 1979), p. 42; cf. S.G. Shanker, 'Computer vision or mechanist myopia?' in S.G. Shanker (ed.), *Philosophy in Britain*, (Croom Helm, London, 1986).

29. J.Z. Young, *Programs of the brain* (Oxford University Press, Oxford, 1978), p. 2; see P.M.S. Hacker, 'Languages, minds and brains', forthcoming.

30. Quoted in Edge, 'Technological metaphor', p. 53 (fn. 13).

31. Michael Mulkay, *Science and the sociology of knowledge*

(George Allen & Unwin, London, 1979), p. 2.

32. Gilbert and Mulkay, *Opening Pandora's Box*, p. vii; cf. Shanker, 'The decline and fall of the Mechanist Metaphor'.

33. For the classic statement on the nature of category mistakes see Gilbert Ryle's *The concept of mind* (Hutchinson, London, 1960), p. 8. For a proper understanding of Wittgenstein's argument, however, see G.P. Baker and P.M.S. Hacker, *Wittgenstein: Meaning and Understanding* (Basil Blackwell, Oxford, 1980).

34. Jamie Fleck, 'Development and establishment in Artificial Intelligence', Chapter 3 in this volume.

35. Richard L. Gregory, *Mind in science* (Penguin, Harmondsworth, 1984), p. 500.

36. Jerry A. Fodor, *The language of thought* (Thomas Crowell, New York, 1975), p. 67.

37. See Hilary Putnam, 'Minds and machines' in *Mind language and reality: philosophical papers volume 2* (Cambridge University Press, Cambridge, 1975) p. 380.

38. See Jeff Coulter, *Rethinking cognitive theory* (Macmillan, London, 1983), pp. 12ff.

39. See Dennett, *Brainstorms*.

40. Mulkay, *Science and the sociology of knowledge*, p. 37.

41. O.K. Bouwsma, 'Wittgenstein notes: 1949', in *Wittgenstein Conversations: 1949–51* (Hackett, 1986) p. 13.

42. Hilary Putnam, 'Reductionism and the nature of psychology', in J. Haugeland (ed.), *Mind design: Philosophy. Psychology. Artificial Intelligence* (MIT Press, Cambridge, Mass., 1985), p. 216.

43. Ragnar Granit, *The purposive brain* (MIT Press, Cambridge, Mass., 1979), p. 69.

44. Aaron Sloman, *The computer revolution in philosophy: philosophy, science and models of mind* (Harvester Press, Brighton, 1978), p. 35.

45. David Marr, 'Artificial Intelligence: a personal view' in J. Haugeland, *Mind design*.

46. Granit, *The purposive brain*, p. 204.

47. David Marr, *Vision: a computational investigation into the human representation and processing of visual information* (W.H. Freeman & Co., San Francisco, 1982), pp. 351, 355; cf. S.G. Shanker, 'Computer vision or mechanist myopia?'.

48. Colin Blakemore, *Mechanics of the mind* (Cambridge University Press, Cambridge, 1977), p. 130.

49. Ibid., p. 91.

50. L. Weiskrantz, 'Experiments on the R.N.S. (Real Nervous System) and monkey memory', *Proceedings of the Royal Society Bulletin*, vol. 171 (1968), p. 337.

51. Morton Hunt, *The universe within* (Harvester Press, Brighton, 1982), p. 18.

52. Cf. Ludwig Wittgenstein, *Remarks on the philosophy of psychology, volume I*, G.E.M. Anscombe and G.H. von Wright (eds), G.E.M. Anscombe (trans.) (Basil Blackwell, Oxford, 1980), § 1063.

53. Ibid.

54. Fodor, *The language of thought*, p. 33.

55. Margaret Boden, *Artificial Intelligence and natural man* (Harvester Press, Brighton, 1979), p. 161.

56. Ibid., p. 85.

57. Ibid.

58. Quoted in Coulter, *Rethinking cognitive theory*, p. 84.

59. Ludwig Wittgenstein, *Remarks on the foundations of mathematics*, 3rd edn (Basil Blackwell, Oxford, 1978), VI § 31.

60. Coulter, *Rethinking cognitive theory*, p. 87.

61. Wittgenstein, *Remarks on the philosophy of psychology*, § 1063.

62. Quoted in Granit, *The purposive brain*, p. 204.

63. Coulter reminds us that 'information' in the theory of thermodynamics refers to 'negative entropy'; cf. his *Rethinking cognitive theory*, p. 29.

64. Norman Malcolm, *Memory and mind* (Cornell University Press, Ithaca, 1977), pp. 217ff.

65. Ibid., p. 219.

66. Andrew Hodges, *Alan Turing: the enigma*, (Burnett Books, London, 1983), p. 251.

67. Warren Weaver, 'Recent contributions to the mathematical theory of communication' in Claude E. Shannon and Warren Weaver, *The mathematical theory of communication* (University of Illinois Press, Urbana, 1963), p. 8.

68. Martin Davis, 'What is a computation' in L.A. Steen (ed.), *Mathematics today* (Springer–Verlag, New York, 1978), p. 241.

69. Ibid.

70. Fred I. Dretske, *Knowledge and the flow of information* (Basil Blackwell, Oxford, 1981), p. 47.

71. Young, *Programs of the brain*, p. 2.

72. Ibid.

73. Jeremy Campbell, *Grammatical man* (Penguin, Harmondsworth, 1982) p. 17.

74. Young, *Programs of the brain*, p. 42.

75. Dretske, *Knowledge and the flow of information*, p. 41.

76. Ibid., pp. 40–1.

77. Ibid., pp. 55–6.

78. Ibid., p. 44.

79. Ibid., p. 54.

80. Cf. Marc de Mey, *The cognitive paradigm* (D. Reidel, Dordrecht, 1982), pp. 19ff.

81. Campbell, *Grammatical man*, p. 113.

82. Ibid., p. 67.

83. Hao Wang, *From mathematics to philosophy* (Routledge & Kegan Paul, London, 1976), p. 91.

84. Campbell, *Grammatical man*, p. 130.

85. Ibid., p. 112.

86. Ibid., p. 123.

87. Steve J. Heims, *John von Neumann and Norbert Wiener: from*

mathematics to the technologies of life and death (MIT Press, Cambridge, Mass., 1982).

88. See ibid., pp. 202 and 476, fn. 14; see also Gregory Bateson, *Steps to an ecology of mind* (Granada Publishing, London, 1978), pp. 17f.

89. cf. John von Neumann, 'The general and logical theory of automata' in James R. Newman (ed.), *The world of mathematics*, vol. 4 (Simon and Schuster, New York, 1956), p. 2089.

90. See John von Neumann, *Collected works*, vol. 5 (Pergamon Press, Oxford, 1963), p. 319.

91. See Claude E. Shannon, 'Computers and automata', *Proceedings of the Institute of Radio Engineers*, vol. 41 (1953), pp. 1234–41; for the highly significant exchange of ideas between Shannon and Turing, see Hodges, *Alan Turing: the enigma*, pp. 250f.

92. Campbell, *Grammatical man*, p. 108.

93. Claude E. Shannon, 'A chess-playing machine' in Newman (ed.), *The world of mathematics*, vol. 4, p. 2124.

94. Cf. Shanker, 'The decline and fall of the Mechanist Metaphor'

95. See Alan Turing, 'Computing machinery and intelligence' in Alan Ross Anderson (ed.), *Minds and machines* (Prentice-Hall, Englewood Cliffs, 1964).

96. Campbell, *Grammatical man*, p. 189.

97. Ibid., p. 193.

98. Cf. S.G. Shanker, 'Computer vision or mechanist myopia?'

99. Young, *Programs of the brain*, p. 62.

100. Weiskrantz, 'Experiments on the R.N.S. and monkey memory', pp. 336–7.

101. Ibid., p. 343.

102. Campbell, *Grammatical man*, p. 197.

103. See Brian Bloomfield, 'Epistemology for knowledge engineers', *Communication and Cognition — Artificial Intelligence*, vol. 3, no. 4 (1986), pp. 305–20; and Shanker, 'The decline and fall of the Mechanist Metaphor' and 'Computer vision or mechanist myopia?'

104. See Granit, *The purposive brain*, Chapter one.

105. Sloman, 'The structure of the state of possible minds', in S. B. Torrance (ed.), *The mind and the machine* (Ellis Horwood, Chichester, 1984). p. 36

106. See Michael Guillen, *Bridges to infinity: the human side of mathematics* (Jeremy P. Tarcher, Boston, 1983); Douglas R. Hofstadter, *Gödel, Escher, Bach: an eternal golden braid* (Penguin, Harmondsworth, 1981); S.G. Shanker, 'Wittgenstein's Remarks on the significance of Gödel's theorem', forthcoming.

107. Hunt, *The universe within*, p. 74.

108. Owen J. Flanagan, Jr., *The science of the mind* (MIT Press, Cambridge, Mass., 1984), p. 227.

109. Ibid.

110. Cf. Shanker, 'The decline and fall of the Mechanist Metaphor' and 'Wittgenstein versus Turing on the nature of Church's Thesis', *Nore Dame Journal of Formal Logic*, forthcoming.

111. Flanagan, *The science of the mind*, p. 229.

112. Ibid., p. 189.
113. Ibid., p. 191.
114. Ibid., p. 229; cf. de Mey, *The cognitive paradigm*, p. 5.
115. Ibid.
116. Cf. L.G. Roberts, 'Machine perception of three-dimensional solids' in J.T. Tippett, D.A. Berkowitz, L.C. Clapp, C.J. Koester and A. Vanderburgh (eds) *Optical and electro-optical information processing*, (MIT Press, Cambridge, Mass., 1965); and Shanker, 'Computer vision or mechanist myopia?'
117. Ibid., p. 189.
118. Ibid., p. 191.
119. Ibid.
120. Margaret Boden, 'Methodological links between AI and other disciplines' in *The mind and the machine*, p. 126.
121. Ibid.
122. Ludwig Wittgenstein, *Lectures and conversations on aesthetics, psychology and religious belief*, Cyril Barrett (ed.) (Basil Blackwell, Oxford, 1978), p. 27.
123. 'The confounded eye', p. 50.
124. Ludwig Wittgenstein, *Culture and value*, G.H. von Wright (ed.), Peter Winch (trans.) (Basil Blackwell, Oxford, 1980) p. 40e.

2

The Culture of Artificial Intelligence

Brian P. Bloomfield

In this chapter I want to tackle the problem of thinking about Artificial Intelligence (AI) as a social phenomenon and to consider what it means to talk about the 'culture of AI'. My principal aim is to show that dealing with the culture of AI does not imply that the investigator must restrict his or her sociological probings to the effect of 'intelligent' computers on individuals or society, but, rather, that he or she must also tackle the social milieu and tradition behind the groups who are the originators and disseminators of the ideas and ways of thinking which characterise AI. In contrast with many existing treatments of the subject, therefore, it is my contention that AI should not be studied as though it were just an abstract body of theory or techniques — as if these could exist in a social vacuum — but is to be understood as a system of knowledge (shared beliefs), a worldview or style of thought (see Fleck, L., 1979; Mannheim, 1936) which is part and parcel of a specific culture shared among a select group of various scientists, mathematicians, and other experts.

The worldview of AI is shared not only among the people who work in universities and research laboratories but also among those increasing numbers who embrace AI in its more popular forms — including those involved in the application of AI techniques in industry and commerce on projects such as the building of expert systems (Pollitzer and Jenkins, 1985). It also includes the growing number of people who are beginning to take on board elements of the ways of thinking and speaking which are the hallmark of AI and seek to incorporate them into their own worldviews — for example, the application of AI techniques in different academic disciplines and in educational training programmes.

59

To get to grips with the culture of AI, I propose to adopt a framework from the sociological study of scientific knowledge: this will not only provide a model of how scientific beliefs are related (though, of course, not determined) to the social circumstances of believers, but will also offer a means to compare and contrast some of the more important alternative viewpoints of AI culture. I should state at the outset that it is not particularly helpful to ask whether AI is actually a science *or* a technology; in fact, AI is an area where the interconnection of theory and technics highlights the general problem which is inherent in any attempt to distinguish between science and technology (see Pinch and Bijker, 1984). For my purposes here, in what follows I shall assume an essential indivisibility between the two; or to put it another way, I will take it as given that any differences between scientific and technological aspects of AI do not significantly affect the thrust of the argument set out here.

It is contended that the body of knowledge which we think of as the science of AI (including beliefs, metaphors, pre-theoretical ideas, ways of thinking and speaking, and cherished hopes) and the social setting of the groups which subscribe to it are intimately intertwined. This is not, however, to seek to reduce one to the other or vice versa; no determinism is implied here. Instead, the interconnection of knowledge and social structure is to be understood in two senses: firstly, the knowledge of AI constitutes a resource by which the people who articulate it both maintain their own group identity and seek to secure status and hegemony among the different competing groups which lay claim to the understanding of the workings of the human mind; and secondly, the nature of the social bonds which bind AI people together (internal), as well as those which regulate their relationships with other disciplines (external), are part of a social context and tradition which has shaped the way that AI has evolved. Thus, social bonds and knowledge, together, comprise the culture of AI: knowledge is not a reflection of social structure, but cannot be understood in isolation from it. While philosophers tend to be concerned with whether the beliefs of the science of AI are *valid*, here I will focus more on the reasons as to why those beliefs are believed. Nevertheless, in showing the connections between beliefs and the social life or culture of a group, one inevitably exposes the particularity of its outlook: that is, one evaluates and delimits the context of that outlook.

My first task will be to indicate the range of perspectives which — in one way or another — directly address or impinge upon the question of AI and culture, and briefly outline some of the flaws which I consider to be inherent in many of these existing treatments. In the following section I will outline the model which will provide the basis for the approach taken here and will then apply it to the case of AI. In the final section I will deal with two very different evaluations of AI culture, those of Turkle and Weizenbaum, and contrast them with the perspective developed here.

2.1 UTOPIA AND DYSTOPIA, THINKING MACHINES AND TOTAL SURVEILLANCE

There exists a near surfeit of literature discussing the shape and social implications of future developments in computing and AI; these have been written by both technological optimists and pessimists. On the one hand, we have those who opine the great emancipatory potential of thinking machines; on the other are those who perceive sinister possibilities for a drift toward a totalitarian society, a state of total surveillance founded on advanced information systems. (For a categorisation of some of the main themes, see Fleck, J., 1984.) It is my contention that the questions which are commmonly asked about computing and AI — such as whether the possibility of creating artificially intelligent machines augers good or ill for society — are very much easier to pose than they are to answer in any really satisfactory sense. Indeed, one could even argue that such questions cannot be answered because they are so general as to leave too much open to speculation. Moreover, a dominant theme which has tended to run through much of the discussion has been predicated on an implicit technological determinism — of which optimists and pessimists are often equally guilty — in which it is assumed that technology is the outcome of an immanent logic of development that *impacts on* society and causes (for good or ill) social change.

By 'social impact' we mean those effects which the technology will have — directly or indirectly — on the society, community, group, or individual; what the social or cultural changes resulting from the technology may be; and what the likely significance of these changes will be for society as a whole. (Gurstein 1985: 655)

To take another example, Boden (1984: 60) expresses a singular confidence in the inevitable onward march of AI:

In the field of artificial intelligence several core research areas are likely to make solid progress within the next decade. Each of these is already being worked on in various countries, and progress does not depend upon the success of Japan's ambitious fifth generation project (though it might be accelerated by associated hardware and software developments).

Furthermore, though she is sensitive to the social implications of the new developments and for the need to introduce them carefully, that we should start thinking now about what the 'optimal social arrangements might be for a post-industrial society' (Boden 1984: 64), her emphasis lies very much on how society might be adapted to fit in with the inevitable changes.

Another strand of the literature does not address questions of potential social impact so much as the significance of AI in terms of existing knowledge, if not the wider development of science and society, and even Western civilization itself. Here we find arguments that range from Weizenbaum's (1976) thesis that computing is closely linked to the increasing reign of instrumental reason in modern industrial societies; through Shallis's more ethically orientated argument that modern technologies, particularly those of computing and AI, effectively constitute the devil incarnate and offer us false gods, that they are 'attempting to play God by redefining nature in their own terms' (Shallis 1984: 80); to the idea that the development of the science of Artificial Intelligence constitutes the emergence of a revolutionary new paradigm engulfing such areas as neurophysiology, psychology and the philosophy of mind. (See Bolter, 1985; Gardner, 1985; Hofstadter, 1978; Turkle, 1984.)

Weizenbaum's view on computers and the role of instrumental rationality is a fairly pessimistic one: '(computing) is . . . an instrument pressed into the service of rationalizing, supporting, and sustaining the most conservative, indeed, reactionary, ideological components of the current *Zeitgeist*' (Weizenbaum 1976: 250). Others, however, foresee a more revolutionary or emancipatory potential in the developments fostered by AI:

> AI is likely, by providing a public and universally accessible form of rationality, to have the effect of stripping away much of the mystique which currently shrouds expertise and decision making in the public sphere and open these up to a more general discussion concerning the assumptions and goals which underlie them (Gurstein 1985: 670).

Another optimist, Bolter, considers AI as a source of unification for the arts and sciences — as if it had the potential to form an interdisciplinary cement or new worldview capable of healing the rift between the 'two cultures' described by C.P. Snow (Snow, 1959).

> My premise is that technology is as much a part of classical and Western culture as philosophy and science and that these 'high' and 'lowly' expressions of culture are closely related. It makes sense to examine Plato and pottery *together* in order to understand the Greek world, Descartes and the mechanical clock together in order to understand Europe in the seventeenth and eighteenth centuries. In the same way, it makes sense to regard the computer as a technological paradigm for the science, the philosophy, even the art of the coming generation. Perhaps from this premise we can establish a much-needed dialogue among scientists, engineers, and humanists (Bolter 1984: xvi).

In a somewhat similar vein, Gardner borrows a Foucaultian metaphor (Foucault, 1970) to suggest that AI is but one of the 'cognitive sciences' alongside phsychology, philosophy, anthropology, linguistics, and neuroscience which together constitute the emergence of a new *episteme* (Gardner 1985: 291).

While it would be possible to raise many objections to these perspectives on the nature of AI, for my purposes here it is sufficient to note the following: firstly, the premature foreclosure or restriction of sociology to questions of social impact; and secondly, the way in which social impacts are conceived in deterministic terms. I will deal with each of these points in turn.

2.1.1 The restriction of sociology

Optimists portray a rather abstract picture of AI as a positive, evolutionary development but tell us little or nothing about the specific culture of the exponents of AI and its relationship to the beliefs which they share. Gardner, for instance, invokes the concept of a new *episteme* rather selectively: he ignores the other side of the Foucaultian knowledge equation — namely, power, and in particular the social structures which are the condition for the emergence of an *episteme* and which are in turn legitimated by it. A similar shortcoming pervades one of the most detailed empirical studies of the culture of AI and computing as described in Sherry Turkle's (1984) book *The second self*. Turkle writes from the perspective of a participant observer seeking to present an account of the range of relationships between people and computers. Her fieldwork brought her into contact with several different groups of people who have an intense involvement with computers — including, home computer buffs, 'hackers' (sometimes known as compulsive programmers), and AI researchers. While claiming to pursue an ethnographic study of AI in the spirit of the sociology of science, Turkle does not examine the connection between the social or cultural setting of AI and the content of its knowledge claims. The result is that Turkle's account might be judged a rather pale reflection of the sort of study now usually aspired to elsewhere among sociologists of science. We receive little or no information about the inherent competition and internal conflicts within and between its various groupings. (For a contrasting view on this particular point see Fleck, J., this volume.) Similarly, in spite of her apparent close contact with these people we are told nothing about the processes of negotiation by which its researchers arrive at interpretations concerning matters such as the usage of concepts or the status and significance of the behaviour of computer programs. (For other criticisms of Turkle's sociological method, see Linn, 1985.)

The foreclosure of sociology also arises with respect to those studies concerned with the social impact of AI, which have similarly ignored the social processes which have shaped and indeed constituted the very products of this area of computing. While purporting to offer an understanding of the social implications of AI, they have invariably omitted reference to arguments that arguably laid the ghost of technological determinism to rest some time ago.

We do not ask for the influence or effect of technology on the human individuals. For they are themselves an integral part and factor of technology, not only as the men who invent or attend to machinery but also as the social groups which direct its application and utilization. Technology, as a mode of production, as the totality of instruments, devices and contrivances which characterize the machine age is thus at the same time a mode of organizing and perpetuating (or changing) social relationships, a manifestation of prevalent thought and behaviour patterns, an instrument for control and domination (Marcuse 1978 (orig. 1941): 138–9).

The restriction of sociology has also been in evidence in studies which have been concerned with mainstream computing — though not to the same degree. More specifically, though many studies of computing in organisations take the technology rather as a given, they do not necessarily conceive its introduction as a mere social or organisational impact. Rather, computing is seen, for example, as a power resource by which different groups within an organisation might seek to further their own particular positions (For a survey of the rather large literature on computing and organisations, see Kling, 1980). The implicit (or even explicit) restriction on sociology which is evident in studies of AI is not accidental: for in fact the disciplines which deal with the study of culture — primarily anthropology and sociology — are not accorded any significant role within AI. As an example of this narrow focus we may note the explicitly restricted role of anthropology within the 'new *episteme*' described by Gardner. Thus, for example, Suchman (1986) draws attention to Gardner's characterisation of the central tenets of cognitive science in which he includes the methodological decision to de-emphasise certain emotional, historical, and cultural factors which would complicate its project. On this account Suchman argues that the actual influence of anthropological ideas on cognitive science is marginal. Thus, if AI can explain human knowledge and thought without recourse to sociology and anthropology, then neither do these disciplines have any real contribution to make to our understanding of how AI — seemingly the latest stage in the development of human knowledge — has itself evolved. (Later we shall see that exponents of AI do sometimes seek recourse to sociological views of science; but this is primarily to explain the antagonism of other disciplines, not to understand the details of their own.)

The exclusion of sociological questions from any serious examination of AI reflects an epistemological bias that has a long history in the philosophy of science. The conventional wisdom once dominant among philosophers of science was that sociology had no contribution to make to the understanding of the content (sometimes described as the *internal* history) of scientific knowledge. Though allowing the possibility that social factors might play a role in shaping the direction or pace of development of knowledge (*external* history), these were thought to have no bearing on its internal rationality. It was once held that scientific beliefs were adhered to simply because they were true; and on this view it was further believed that developments in scientific knowledge could be deduced from methodological rules alone. This model of science no longer retains validity: recent work in the sociology of science has painted a far more 'realistic' picture of how scientific beliefs are embedded within a social context of legitimation which includes processes of bargaining and negotiation concerning matters such as the interpretation of experimental findings or the choice between rival theories.[1] In short, it is now accepted that facts do not speak for themselves but are mediated by the collective judgement of scientists: they are social constructions (see Barnes, 1982; Barnes and Shapin, 1979; Collins, 1985; Knorr-Cetina and Mulkay, 1983; Shapin, 1982).

More recently there has been a move to extend the sociology of science toward examining the social construction of technology. (For a review of this work, see Pinch and Bijker, 1984). While the conventional view has it that technology is politically neutral or value-free, that it can be used for either good or ill and develops according to an inner logic, sociologists of technology aim to show how social factors can not only shape the direction of development but also the internal details of particular technological artefacts. To take one pertinent example, let us consider the early development of computers during the Second World War. While some might well accept that the circumstances of the period helped to stimulate particular innovations in computing (for example, in order to improve the efficiency and potency of war-related technologies), social factors did not influence the internal details of the particular machines that were developed. However, Kraft's (1979) analysis of computing offers us a rather different picture: for instance, he shows how the prevailing social attitudes toward the work of women and the resultant division of labour had an influence on the specific pattern of development of computing in the 1940s and 1950s. Kraft describes how the project to build the first operational computer — ENIAC — was conceived largely as a task of electrical engineering: the emphasis was on getting the hardware up and running, while providing it with instructions (programming) was viewed rather as a last minute clerical detail — in short, women's work. However, the women who were hired to carry out this task proved just how complex and important it actually was: 'Programming required familiarity with the machine's electrical logic, its physical structure, and its mechanical operation. The women learned by crawling around the ENIAC's massive frame, locating burnt-out vacuum tubes, shorted connections, and other nonclerical "bugs"' (Kraft 1979: 4). As a result of the ENIAC project, the attitude of the electrical engineers towards programming changed until it was eventually no longer defined as women's work; moreover, in time, men effectively displaced women from this section of the industry; and, of course, once the engineers gradually realised just how important programming was, the way was opened for new forms of computing. (For a detailed analysis of the development of another early computer — *Whirlwind I* — see Redmond and Smith, 1980.[2])

2.1.2 Social impacts

The second objection to the conventional views on AI is the implicit assumption that there is a one-way diffusion of scientific knowledge (e.g. from AI research centres) which impacts on society and causes social change. In a sense, the determinism underpinning this assumption is a corollary of the restriction on sociology: if science and technology unfold in accordance with an inner logic, so also does their effect on society. The problem here is that no allowance is made for the possibility that in the world outside research centres, the use of AI tools and techniques might not take the form of a direct 'impact' so much as a local elaboration which is wedded to rather different goals and interests. In contrast, it is perfectly plausible to expect that the various tenets, metaphors, or ways of thinking which are indigenous to AI, do not determine social impacts but, rather, undergo metamorphosis as they become woven into the texture of social life within different organisational or cultural contexts. An example of this possibility is noted by Boden in relation to contrasting mechanistic and non-mechanistic views of humanity which might stem from different public interpretations of AI (Boden 1984: 64). Indeed, people are not necessarily passive recipients of AI: in figurative terms, we might say that some actively enter into, or build up, their own local 'outposts' of AI culture.

2.2 SCIENTIFIC KNOWLEDGE:
THOUGHT COLLECTIVE AND THOUGHT STYLE

Woolgar (1985) argues that a sociology of intelligent machines should take as its topic the dichotomies and distinctions sustained in the discourse of AI. However, it is also important to examine the concepts and ways of thinking of the AI community in relation to the tradition in which it stands and the social contexts in which it is located. Thus, the discourse of AI does not exist in isolation from the nature of its surrounding social milieu, including the social bonds among AI researchers themselves and between them and those in other areas — indeed, it is shaped by the interactions among these groups. Nor can we allow any *a priori* distinction between thought and feeling to overemphasise the role of discourse at the expense of our understanding of the part that hopes and wishes might play in sustaining the particular social groups which embrace AI. To get to grips with matters such as these I will now consider the work of Ludwik Fleck.

In the 1930s Fleck developed a model of science which revolves around the idea that scientific knowledge is the product of a 'thought collective' comprising a group of people who maintain an intellectual interaction exchanging ideas or cognitions. Over time, the thought collective provides a carrier for the stock of knowledge in a particular field; Fleck described this as the 'thought style' of the collective. He argued that the development of knowledge always takes place within a thought style:

> The individual within the collective is never, or hardly ever, conscious of the prevailing thought style, which almost always exerts an absolutely compulsive force upon his thinking and with which it is not possible to be at variance (Fleck, L. 1979: 41).

Furthermore, knowledge is characterised by two forms of connections: firstly, there are 'active' connections which evolve within the thought style and as such are not freely chosen but are part of the tradition or culture shared by the thought collective; and secondly, there are 'passive' connections which are imposed by nature and effectively experienced as objective reality. To give an example, the choice of 16 for the atomic weight of oxygen is conventional in character and represents an active connection; but once fixed in this way the choice determines the fact that the atomic weight of hydrogen will be measured as 1.008. In Fleck's words: 'This means that the ratio of the two weights is a passive element of knowledge' (Fleck, L. 1979: 83).

Within any given thought collective we find mutually interdependent 'esoteric' and 'exoteric' circles:

> The general structure of a thought collective consists of both a small esoteric circle and a larger exoteric circle, each consisting of members belonging to the thought collective and forming around any work of the mind, such as a dogma of faith, a scientific idea, or an artistic musing. A thought collective consists of many such intersecting circles (Fleck, L. 1979: 105).

Their mutual interdependence arises in so far as the exoteric masses trust the esoteric elite for the interpretation of the knowledge within the thought style (the masses' contact with the products of the thought style is mediated esoterically); while, conversely, the esoteric circle is dependent upon the opinion of the exoteric circle.

> Trust in the initiated, their dependence upon public opinion, intellectual solidarity between equals in the service of the

same idea, are parallel social forces which create a special shared mood and, to an ever-increasing extent, impart solidity and conformity of style to these thought structures (Fleck, L. 1979: 106).

However, aside from the force of public opinion, many research areas of course rely upon funding from external sources such as government bodies or industrial/commercial organisations, and the greater the exoteric support they enjoy from within these sources, the more likely their continued funding.

The dynamic of the power relationship between the esoteric and the exoteric circles leaves its mark on the products of the thought style. In particular, the esoteric circle tends to conceal research difficulties from the exoteric circle. 'If the elite enjoys the stronger position, it will endeavour to maintain distance and to isolate itself from the crowd. Then secretiveness and dogmatism dominate the life of the thought collective' (Fleck, L. 1979: 106). As an historical example of this tendency, Fleck refers to the way in which the public was kept in the dark about the anomalous status of the orbit of Mercury in relation to the Newtonian framework.

Finally, we may note two processes in the development of knowledge which will be of particular relevance later when the popular or exoteric role of AI is referred to. Firstly, we find a process in which popular exoteric knowledge in the shape of vague 'proto-ideas' from the background culture — e.g. the long-held notion that syphilis was associated with bad blood — gradually become refined and embedded within the theories of the esoteric circle — such as the modern epidemiological model of syphilis and its detection by the Wassermann blood test. (For a detailed analysis of the genesis of another proto-idea, within the context of logic and the law, see Leith, this volume.) Secondly, there is a movement of ideas from the esoteric to the exoteric circle which is not one of simple linear diffusion but is a process involving the 'popularisation' of esoteric knowledge; this becomes interpreted in line with the specificities and commonsense understandings which characterise the cultural contexts of the particular exoteric group involved (see de Vries and Harbers, 1985). To give a topical example of this process we may consider the differing opinions emanating from various religious groupings in relation to the AIDS (Acquired Immune Deficiency Syndrome) 'epidemic'. Drawing on the clinical findings of the esoteric circles of bacteriology and epidemiology, some groups conclude that the disease is a form of divine retribution for unnatural sexual practices; on this view such indulgences should be expressly forbidden. Others, however, take the more liberal view that in order to minimise the statistical chances of contracting the disease, people should be more conservative in their choice of partners and avoid promiscuity. In other words, the same esoteric findings are disparately interpreted to fit in with the commonsense outlooks of different exoteric groups. This alternative view of the diffusion of scientific ideas urges that the social implications of AI should be viewed in a different light to conventional treatments: the metaphors and ways of thinking of AI do not impact on an unwitting public; rather, different groups enter the thought collective of AI bringing with them their own particular constellations of goals, interests and interpretations.

2.3 THE THOUGHT COLLECTIVE OF AI

At the outset, it is worth briefly mentioning that the use of Fleck's model of science does not mean that I wish to underplay the heterogeneity of the thought collective of AI: indeed, Fleck explicitly stated that individuals may be members of several exoteric and esoteric circles. However, just as there is diversity, so also there is a commonality in AI; a kind of identity distinguishes AI researchers and exoteric laypeople from outsiders and it is this unity which will be the focus of attention here.

Beginning with the background culture within which AI is embedded and with the exoteric circles which both thrive on and in turn nourish it, we may usefully refer to several features of particular interest. Firstly, there is the public or conventional view of those experts who deal with computing machines and the high esteem in which they tend to be held. In fact, computing plays an increasingly important role in modern industrial societies: for example, in commerce and industry, in the functions of government (e.g. defence or taxation), in public utilities and in communications, complex technological and administrative systems have evolved which could conceivably collapse overnight without computer support. As the importance, power, and sophistication of modern computer systems is ever more widely acknowledged, those who program and control them have come to be highly regarded and accordingly rewarded in the labour market. Secondly, computing is an endeavour which tends to be particularly shrouded by myths and misconceptions which can be only partly explained by the obvious fact that so few people have a grasp of its essentials. Just as chess players are often automatically thought to be particularly clever or intelligent *per se*, it seems that computer professionals are also often viewed as rather cognisant people because they are considered experts in a technology which — though increasingly commonplace — is essentially opaque to the majority. While opacity is a distinguishing feature of many other areas of science and technology, the myths surrounding computing may stem less from the fact that it is an opaque esoteric subject and more from the way in which it can be seen to blur the boundary between people and machines (Turkle, 1984). To be sure, most people do not understand the workings of a television set or how to program their video cassette recorders properly, but then they do not usually believe that these machines can have intelligence. The public myths about computing and AI are also no doubt due to the ways in which computers are often depicted in the mass media — e.g. as an abstract source of wisdom, or as a mechanical brain.

A third factor of interest is the emergence, particularly in recent years, of a strong exoteric following for work in computing and AI; for example, books such as Hofstadter's (1978) *Godel, Escher, Bach: an eternal golden braid* have enjoyed great popular success. Moreover, there has been a tremendous burgeoning of popular and semi-popular literature on computing and AI in newspapers and scientific magazines. Also, powerful commercial interests have led to a great expansion in the use of AI-related technologies in business and other organisations; while among academics, new journals (representing a confluence of academic and commercial interests) have sprung up to synthesise work in AI with that emanating from disciplines which have been traditionally non-computer orientated (e.g. history).

Fourthly, researchers in many other disciplines might be said to have jumped on the AI bandwagon. This has taken two forms: some have proposed projects to examine the implications of AI in their own particular field, while others have redefined projects from within existing research programmes in terms of goals and objectives that overlap with those of AI. In each case it has been thought that AI-oriented projects are likely to have an increased chance of funding in an environment in which resources are generally rather tight.

The exoteric circles of AI are important in two respects. Firstly, they constitute expressions of AI culture in their own right; and secondly, they can influence the fortunes of the esoteric circle. While it is difficult to assess the impact of the foregoing diverse factors on the esoteric circle of AI, they arguably play a role in strengthening its ethos as well as providing a useful source of legitimation. This source is not just important for academic researchers who wish to justify their projects to a wider audience or to funding bodies; for it can also play a role in shaping the self-perception of the increasing number of up-and-coming AI researchers when making career choices or when needing to justify their work to themselves (note that in terms of age structure, AI is a relatively youthful field). Thus, when new researchers begin to assimilate the culture of their field, they often have to take certain things on trust, to believe; supervisors can usually play an important supportive role in this regard but the broader societal view of the field may be no less important.

Having been given much buoyancy by the rhetoric and hype stemming from commercial interests and the media, the AI balloon has gained an apparently unstoppable momentum; and coupled with the mystique that surrounds the whole enterprise, the resultant potion has provided a fertile ground for the formulation of AI research projects — thus, Athanasiou argues that the mystification of work in AI is 'good for business' (Athanasiou 1985: 15). In this regard, the public debate about the 5th Generation has been such that the stated purposes of the projects in different countries have gone largely unchallenged.[3] Instead of a discussion about who will develop what form of machines, for use by whom, there would appear to have been a general acquiescence in assuming that the 5th Generation was an inevitability. This point is all the more crucial when one realises that a number of 5th Generation projects were not financed by new sources of funding but, rather, caused the diversion of monies from other research areas. What is of particular importance here is that the increase in funds for AI work has not been based on a broad scientific consensus about significant

73

progress in AI but is arguably a reflection of the more general notion or idea that AI is a field whose time has now come. In part the latter is due to the measure of success in which AI has managed to secure a monopoly over the 'means of orientation' concerning intelligence, whether human or machine (Elias 1982b: 37; see also Fleck, J., this volume). Of course, behind the euphoria powerful interests in commerce, industry and the military have been at work (Burnett, 1985; Dreyfus and Dreyfus, 1986; Solomonides and Levidow, 1985). In the British context this state of affairs is particularly noteworthy in view of the recent history of AI research during the 1970s. As a result of the 'Lighthill report' (Lighthill, 1973), research funds for work in AI were drastically curtailed: now, however, it seems that AI has returned with a vengeance.

Turning now to the esoteric circle of AI, there is of course (as indicated earlier) a diversity in the social settings within which AI-related projects and practices take place. Spanning different continents and societies, AI techniques are employed in industry, commerce, state organisations, and universities — each of which must impart its own specific stamp on the activity in question. Thus, for example, AI programmers in commerce and industry often tend to work as part of a team, whereas in academia they tend to develop programs on their own, and this diversity in the social organisation of work has import for the respective styles of programming which emerge. (See Kraft, 1979; Weinberg, 1971.)

Deciding just who is part of the esoteric circle of AI is a difficult job and is related to the problem of trying to define what actually constitutes work in AI: some appear to believe that the very use of particular programming languages such as PROLOG constitutes the definitive character of AI work (Lehnert, 1985); others feel that it is determined by the nature of the application in hand — e.g. modelling some cognitive function. Moreover, there exists a good deal of variation in the beliefs among AI people about the nature of the subject and where it is going. Some follow the 'weak' AI thesis, for instance, in which it is believed that in principle, it is possible to construct computer programs that can emulate human skills: this contrasts with the 'strong' thesis in which it is believed that simulations of thought do not merely mimic human skills but actually are thought. Another pertinent polarisation in this regard is that between so-called 'neats' and 'scruffies' (Durham,

1986). The 'neats' are those AI researchers who emphasise the need to put the subject on a more formal footing; with a firm faith in formal logic, they believe, for example, that it is necessary to provide a tight specification for all knowledge-based systems. In contrast, 'scruffies' avoid formal logic and prefer to develop their own particular knowledge representations; they believe that human intelligence is too multifaceted to be captured by a single elegant piece of mathematics.

The organisational environments within which AI was fostered — and in particular the military dimension during and after the Second World War — can be seen as especially important in nourishing the energies and singular sense of purpose among some of its chief exponents. The early pioneers at centres such as the Massachusetts Institute of Technology (MIT) and the Rand Corporation were effectively 'given their heads' and encouraged to 'think big'. According to McCorduck, the Rand philosophy at that time can be summarised by the phrase: 'Here's a bag of money, go off and spend it in the best interests of the Air Force,' (McCorduck 1979: 117). In broader terms, the effect of the military setting in relation to the development of computing and associated technologies is forcefully captured by Noble:

> For the war-related work of the scientists and engineers, their ability to invent, experiment, and theorize, derived not only from the power of their intellects and imaginations but from the power of their patrons as well. It was the political and military power of established institutions which rendered their often fantastic ideas viable and their unwieldy and expensive inventions practical. And it was the assumption of such social power that guided the technologists in their work, giving them the specifications to satisfy and the confidence to dream. And while the new technologies and theories, formally deterministic and intrinsically compelling, compounded the traditional compulsions and enthusiasms of the scientific community, they reflected also the needs of those in command, adding immeasurably to their power to control, and fuelling their own delusions of omnipotence. Thus, the worlds of science and power, having converged in spirit and deed, gave rise together to a shared worldview of total control (Noble 1984: 55–6).

For a more detailed picture it is useful to consider Turkle's study of the AI researchers at MIT which, despite its methodological shortcomings, gives us an insight into the ethos and *esprit de corps* of members of the esoteric circle of AI. (For a detailed mapping of the genesis and development of the main esoteric groupings within AI, see Fleck, J., this volume.) Turkle describes the AI community at MIT as intellectual imperialists: with an encultured habit of thinking on a global scale, they see themselves as being opposed by reactionaries (into whose academic domain they trespass) who hold to more traditional beliefs about the nature of the mind.

Being in a colonizing discipline first demands and then encourages an attitude that might be called intellectual hubris. You need intellectual principles that are universal enough to give you the feeling that you have something to say about everything. The AI community had this in their idea of program. Furthermore, since you cannot master all the disciplines that you have designs on, you need confidence that your knowledge makes the 'traditional wisdom' of these fields unworthy of serious consideration. Here too, the AI scientist feels that seeing things through a computational prism so fundamentally changes the rules of the game in the social and behavioural sciences that everything that came before is relegated to a period of intellectual immaturity. And finally you have to feel that nothing is beyond your intellectual reach if you are smart enough (Turkle 1984: 260).

Writing from a very different (Marxist) perspective, Athanasiou's account of AI notes rather similar features:

The culture of AI is imperialist and seeks to expand the kingdom of the machine . . . The AI community is well organised and well funded, and its culture fits its dreams: it has high priests, its greedy businessmen, its canny politicians. The US Department of Defense is behind it all the way. And like the communists of old, AI scientists believe in their revolution; the old myths of tragic hubris don't trouble them at all (Athanasiou 1985: 13-14).

Turkle contends that AI people aim to challenge the concept of truth and see themselves as a new breed of philosopher. 'These global aspirations are an expression of the pervasive intellectual hubris of AI, part of its sense of being an enterprise of mythic proportions' (Turkle 1984: 269). Within this enterprise, not only the creation of artificial intelligence but also the building of robots is of central importance. Moreover, in a very suggestive footnote, Turkle tells us that a number of the AI people at MIT (including Marvin Minsky, Joel Moses, and Gerald Sussman), as well as previous American scientists (such as John von Neumann and Norbert Wiener), have grown up with a family tradition that they are descendents of a certain Rabbi Loew who was 'the creator of the Golem, a human-like figure made of clay into whom God's name breathed life' (Turkle 1984: 270; for further details, see McCorduck 1979: 12-13). In other words, these scholars at MIT share a rather unique vision of their mission and sense of purpose; and moreover, this sharing is not just an intellectual matter, for it constitutes part of the social bonds — the tradition — which helps to bind them together as members of the thought collective. In fact, the desire to build a robot or a mechanical man has long been a dream in Western thought and must be considered as part of the deeper cultural roots of AI. (For a discussion of the historical background, see Fleck, J., 1984.)

On a point of historical comparison it is worth drawing attention to the similarities between the AI grouping at MIT and the System Dynamics Group which is also located there. System Dynamics (an offshoot of systems analysis) is a policy oriented technique for building computer simulation models of social systems which was developed by Jay W. Forrester — a distinguished engineer who was project leader during the construction of the *Whirlwind I* computer during the 1940s and 1950s. (For seminal works in System Dynamics, see Forrester, 1961, 1969, 1971). Two points of similarity which are particularly noteworthy are: firstly, System Dynamics shares some of the same intellectual roots as AI — such as systems engineering, operational research and cybernetics, and the development and application of these fields within a military context during and immediately after the Second World War; and secondly, during its development the System Dynamics Group shared a strong sense of being part of a unique mission embracing a new philosophy which was believed to be superior to conventional thinking. Elsewhere (Bloomfield, 1986a) I have argued that MIT provided fertile ground for the emergence of System

Dynamics; and as for the significance of its parallel with AI, perhaps the exclusive and elitist nature of that institution is the most telling feature. Of course the AI group at MIT is only one part of the esoteric circle, however. Whether the organisational settings of the other leading groups — such as the Stanford Research Institute, Stanford University, and Carnegie-Mellon — have provided similar nurturing environments is a moot point; certainly, each centre has tended to evolve its own style of AI research (McCorduck, 1979).

2.3.1 Relationship with other disciplines

The foregoing discussion has focused mainly on the internal relationships of the thought collective of AI. The next task is to look at its external relationships with other academic areas and traditions, for these are also important in shaping the beliefs of the collective and maintaining its unique sense of purpose. System Dynamics received a fairly hostile academic reception — achieving most notoriety during the infamous 'limits to growth debate' of the early 1970s — but was given strong exoteric support (Bloomfield, 1986a). The question that immediately arises is whether AI has experienced a similar pattern of acceptance and rejection. I have already dealt with the strong exoteric following of AI but must now deal with the opposition.

In fact, it is widely accepted that AI is a controversial area, providing a focus for bitter and acrimonious exchanges among academics in a number of different disciplines: it is a battle that is conducted on both esoteric and exoteric ground. (For an esoteric phenomenologically based critique from the arch-rival of AI, see Dreyfus, 1972; for an exoteric critique which was included in the annual BBC Reith Lectures, see Searle, 1985.) As with other notorious controversies in science, the charges against AI are sometimes loaded with moral indignation; allegations of cheating are not uncommon. In part the animosity toward AI is to be explained by the fact that projected developments in advanced computing appear to entrain such important consequences for so many areas of science and life in general. It is no exaggeration to say that AI goes to the heart of important philosophical and cultural concerns which have preoccupied scholars for centuries. However, the hostility is also due to the fact that AI is truly a Promethean science: it holds out the prospect of recreating the human mind in the image of an assemblage of logic circuits. Alternatively, depending on one's point of view, AI is a dark art whose practices seem to threaten a mechanisation of the human spirit — offering a dystopia in which people will become slaves to a pale reflection of themselves as embodied in 'intelligent' machines. The latter sentiment is almost encouraged by AI people themselves:

> Automatic creation of knowledge has unpredictable effects. When a machine can use up all knowledge we have given it, and use it systematically in ways that we cannot, and can make inferences more deeply than we can (because it is not limited, as we are, by our evolutionary legacy of about 4 items that we can attend to simultaneously) what will happen? (Feigenbaum and McCorduck, 1984: 322-3).

Another example of similarly divisive rhetoric is to be found in the assertion (articulated with seemingly provocative intention by Minsky) that the human brain is a 'meat machine' (McCorduck, 1979).

Aside from philosophical and humanist critiques, more recent voices of criticism have spoken out against the overt commercialism of the applications-oriented strands of AI (e.g. expert systems which are sometimes viewed as the engineering side of AI; for a discussion of legal applications, see Leith, 1986), not to mention the crucial role of AI within the Strategic Computing Initiative (SCI) of the American military-industrial complex. In the case of the former, one critique suggests that AI 'engineers' are more interested in becoming rich than playing God (a charge levied against the AI 'scientists').

> The engineers like to bask in the reflected glory of the AI scientists, but they tend to be practical men, well schooled in the priorities of economic society. They too worship at the church of machine intelligence, but only on Sundays. During the week, they work the rich lodes of 'expert systems' technology, building systems without claim to consciousness, but able to simulate human skills in economically significant knowledge-based occupations (Athanasiou 1985: 14).

With the SCI, the arguments railed against AI stem from the

perceived danger in allowing complex and enormously powerful defence systems to be automatically controlled by machines. The argument here is advanced on two fronts: firstly, that computer systems of the required complexity cannot be guaranteed infallible; and secondly, that computers are simply not capable of the degree of sophistication and flexibility in judgement available to people (Computer Professionals For Social Responsibility, 1984; Dreyfus and Dreyfus, 1986).

Apart from the overt arrogance of the AI project, it is also clear that some of the antagonism that exists between AI and its critics stems from the fact that the different sides in the debate hold fundamentally conflicting or incommensurable world-views. In other words, the heated arguments characteristic of debates concerning AI do not arise so much from semantic differences in terminology, as from the fact that the opponents actually live in *different* worlds and continually 'talk-past' each other. This tension in the social bonds between AI and other areas can be seen as a reflection of the conceptual walls that AI people perceive between themselves and the outside world; that is, the gulf between AI and traditional disciplines is mediated and reinforced by these social divisions. In addition, some of the leading exponents of AI have made reference to Kuhnian style sociology of science (Kuhn, 1962) in order to explain (away) the hostile criticism that has emanated from other areas. Drawing on the notion of paradigms and scientific revolutions, AI is posited as a revolutionary development in the study of the mind which is opposed by the conservative forces (paradigm reactionaries) from within older and more traditional disciplines (see Shanker, this volume). Such rhetoric on the part of the exponents of AI not only serves to account for the reaction of outsiders but also helps to maintain the perceived legitimacy of AI and justifies the gulf between them and their opponents. In short, it is believed that AI does not need to learn any lessons from those areas which have toiled with the nature of knowledge and thought for some considerable time — such as philosophy, anthropology, and sociology: rather, it is the latter which have much to learn from AI. The persuasive ability to explain away the criticisms of outsiders also plays an important role in maintaining the allegiance of the exoteric circle — in particular among those whose only knowledge of the philosophy of science is mediated by the popularisations of the esoteric circle.

2.4 THOUGHT STYLE AND DISCOURSE

The thought style of AI can be usefully viewed from several perspectives. Here I will discuss some general features of the discourse of AI (the common ways of thinking and speaking about the subject) and thereby begin to piece together a picture of some of the underlying assumptions on which the thought style is based. Other, more detailed views of the thought style are developed elsewhere in this volume: these largely deal with the esoteric aspects of the thought style, including such matters as theoretical precursors, intellectual coherence, disciplinary roots, historical continuity and discontinuity, negotiation and development (see Fleck, J.; Leith; and Schopman, this volume). In contrast, the focus here will be on the more general esoteric *and* exoteric aspects.

As a starting point let us consider the air of aspiration that tends to characterise adherents of AI — irrespective of their internal differences. It is rather an understatement to assert that the whole spirit of AI cannot be appreciated in isolation from the belief that significant developments in computing lie 'just around the corner' — indeed, this refrain is a strikingly commonplace feature in both esoteric and exoteric parts of the literature. This hope is to be particularly observed when people in AI reflect on the state of the art in the field: thus, it is believed that significant developments or breakthroughs in AI would be within reach 'if only there was a machine big enough', or 'fast enough', etc. It is this sort of rhetoric which helps to sustain the idea that projected developments such as parallel processing or the 5th Generation signal the qualitative breakthrough in technique that will make the longstanding problems faced by AI simply melt away; it would also appear to dull the memory of past failures (Dreyfus and Dreyfus, 1986). As one commentator puts it: 'The culture of AI encourages a firm, even snide, conviction that it's just a matter of time. It thrives on exaggeration and refuses to examine its own failures,' (Athanasiou 1985: 18). In this context we should not ignore the fact that AI has been promising significant developments since its very inception, and that such promises are intimately bound up with its style of thought.

There are now in the world machines that think, that learn, and that create. Moreover, their ability to do these things is going to increase rapidly until — in the visible future — the range of problems they can handle will be coextensive with the range to which the human mind has been applied. (Simon and Newell, 1958: 8).

The importance of the belief in the universal applicability of the concept of 'program' (Turkle, 1984) is worthy of mention because it provides a touchstone around which all strands of opinion in AI can unite; in so far as it can be said to apply to all things, the program concept bears some resemblance to the concept of 'system' as articulated by cyberneticians and General Systems Theorists — anything can be considered to be a system (Lilienfeld, 1978).[4] The belief in program is allied to a mechanical view of knowledge and thought wherein the followers of the 'strong' AI thesis go so far as to contend that the simulation of thought (e.g. by a program) actually is thought. (For criticisms of this view, see Searle, 1980; Shanker, 1987.) While this might sound somewhat similar to behaviouristic psychology, Bolter points out how many people in AI regard themselves as humanists — in so far as they seek to reveal the *sophistication* of the symbol processing power of the mind. That said, Bolter nevertheless goes on to conclude:

But programmers with their semantic networks and behaviourists with their Skinner boxes agree on this vital point: everything that happens in the mind or the brain is played out according to the rules of a formal system. The rules are finite, and they can someday be specified (Bolter, 1984: 220; see also Shanker, this volume).

At root we have to appreciate the importance of this predicate if we are to make sense of many of the other assertions found in AI: for example, when Minsky tells us that computer science 'makes it possible to manipulate ideas as though they were things' and that 'there is nothing new in the idea that knowledge can be manipulated like a material; we could always rearrange the pages of a book,' (Minsky 1979: 392) or, when he argues (Minsky 1975: 260-2) that there is an overlap between the AI concept of a 'frame' — a formal description of a context used in knowledge representation — and Kuhn's idea of a paradigm. (As Dreyfus (1981:185) points out, whereas a frame is a formal description, a paradigm in the Kuhnian sense is a 'shared concrete case'; paradigms are shared by a community of scientists and cannot be formally articulated.) Similarly, the mechanical approach is evident in Hayes-Roth's (1983) crude attempt to transform the subtleties of creativity that characterise informal mathematics — as described by Lakatos (1976) — into computer programs.

That way of speaking, which imparts intelligent agency to computers, constitutes another crucial element of the discourse of AI. Irrespective of the complaints of philosophers, popular films and science fiction literature have, in a sense, already settled the issue. To take a commonplace exoteric example, people often talk of being beaten by a computer such as a chess-playing program: the contest is viewed as one between a human player and a thinking machine rather than one between people — i.e. the human player and the programmer who wrote the program. The importance of such habits of discourse should not be underestimated; they are not confined to an 'uncritical' public. Indeed, they are also a hallmark of the ways of thinking and speaking of some who are among the esoteric elite of AI. Thus, for example, consider Michie's (Michie and Johnston 1985: 107) report concerning the program *AM* written by Lenat (1977):

This started with an extremely basic set of mathematical concepts and 'wandered around' the problem space looking for more. *It* constructed *on its own* a number of well-known mathematical ideas and even *rediscovered* important theorems, such as Euclid's Unique Factorization Theorem, which states that a composite number can be factorized into primes in only one way. *It* formulated a curious geometric interpretation of Goldbach's Conjecture, to do with the sums of primes, and made one discovery concerning 'maximally divisible numbers' that was entirely original (emphasis added).

What this indicates is that the language habits which appear to attribute self-directed intelligence to computers are not always employed as mere linguistic conveniences that function only as a kind of verbal shorthand. Rather, in both esoteric and exoteric circles they are linked to explicit claims about the intelligent status of computers and to specific interpretations concerning the achievements and status of particular programs.

83

These claims and interpretations are context dependent; conversely, they help to mediate and reinforce the social bonds which cement the thought collective, to distinguish the believers from the non-believers, the insiders from the outsiders.

Woolgar has discussed other notable features of the discourse of AI (Woolgar, 1985). Among them he notes the inherent interpretative flexibility in many of the important terms which are employed in AI. Woolgar argues, for instance, that terms such as 'expertise' or 'intelligence' can be variously interpreted in relation to the claims about particular theories of intelligence or the programs which allegedly embody them; and the upshot of this interpretative flexibility is that consensus about the state of the art is fragmentary and transient. (For a parallel argument concerning the experimental conditions and interpretation of the Turing Test, see Collins, 1986.)

The assumption of technological determinism which forms another building block of the rhetoric of AI can be seen as the other side of the coin of the faith that significant developments lie just ahead. Like members of many other disciplines, people in AI sustain a view of the history of their field which displays a certain element of wish fulfilment (Elias, 1978a): that is, they perceive that history as they might wish it to be rather than as it necessarily has been. As an example we may consider the 5th Generation. For some, the news of the Japanese 5th Generation project was seen as simply part of an unfolding drama in which the development of computing was assumed to be moving inevitably toward the realisation of a machine capable of human intelligence. For those in the exoteric circles who did not care to follow the technical arguments about the steady advance toward machine intelligence, the myth of continuous rapid progress, coupled with the simple and seemingly inescapable fact that computers have shrunk considerably in physical size while at the same time sustaining enormous increases in computational speed and power, might have seemed sufficient to settle the issue. However, a strongly conflicting view of progress in computing has been put forward by Catt (1973) who was himself once a computer design engineer. He describes how commercial pressures give rise to constant incremental changes in computers which are dressed up as significant advances and serve to perpetuate the myth of continual rapid progress. (To give another example, it is known that some computer manufacturers deliberately slow down initial versions of their machines in order to be able to entice purchasers into accepting frequent hardware or software upgradings.)

However, technological determinism has deeper roots than the apparent dynamic of computer hardware: in fact, seeing the 5ᵗʰ Generation project as the logical outcome of current trends in AI actually mirrors a strong strand of technological determinism within computing as a whole and is perhaps conveyed most vividly by the explicit way in which the history of computers is commonly portrayed as a sequence of generations. To be sure, we know that both the hardware and software of computing have undergone tremendous changes since the early days of the 1940s, but the generational view of computing says much more. *Inter alia*, it implies that the history of computing has been — and indeed will be — punctuated by successive revolutionary advances, each becoming embodied in a new generation of machines. This generational view of computing is in keeping with much of the conventional wisdom on the progress of science and technology in general, which assumes an onward advance in accordance with an immanent logic of development. Thus, we have an image of a persistent rising curve of technological development — moving from the substitution of human muscle power by machines to the substitution of cognitive capacities (such as ability to do arithmetic) by the early computing devices; while the 5ᵗʰ Generation computers will allegedly duplicate the higher mental functions such as reason, language, and judgement. In addition, while the 5ᵗʰ Generation is still arguably very much a gleam in the computer engineer's eye, there is already much talk in certain circles of possible designs for the 6ᵗʰ Generation, including the idea of building analog bio-chips to simulate the operation of neural networks (*Machine Intelligence News*, 1985). (With this in mind, it is worth noting a recent suggestion in *Computing* (1986) that the Sun may have prematurely set on the Japanese 5ᵗʰ Generation project.)

Just as the hardware of computing is seen in progressive developmental steps, so also is the evolution of software — from machine code to high-level languages, from data processing to knowledge processing, and from procedural to declarative programming. Currently within AI there is a major shift towards logic programming and PROLOG (PROgramming in LOGic) has come to be seen as the language *par excellence* of 5ᵗʰ Generation computing. It is no secret that many PROLOG programmers tend to look down on the older and less sophisticated languages such as COBOL and FORTRAN. In addition, there is an ongoing debate between the exponents of PROLOG and those who use the other major language of AI — namely, LISP — as to which is the best language for AI work. (In strictly historical terms, LISP is an old language — it was devised in the 1950s and 1960s — but is still the dominant one in terms of AI applications.) Thus, the abstract hierarchy of computer programming languages is matched by a status hierarchy: the apparent superiority of the higher levels in terms of sophistication, complexity, and power, encourages feelings of self-esteem among those who use them and reinforces the belief that they are at the cutting edge of research. (For a discussion of the problems engendered by the ideological involvement with PROLOG, see Lehnert, 1985; Leith, this volume.)

Though the generational view of computing undoubtedly provides a neat and tidy picture, it is for that very reason a problematic one. A number of arguments can be advanced to substantiate this point and by noting them here we can better understand the implications of holding the generational view. One stems from the fact that in practice, clear-cut boundaries do not exist between the machines of one putative generation and those which have either preceded or followed it. Another argument is that, depending on one's focus of attention, the history of computing can be interpreted in various ways. Table 1 shows three possible schemes: one — due to Feigenbaum and McCorduck (1984) — centres mainly on developments in hardware; a second focuses on software; while the third concentrates on changes of application or problem domain. A third argument against the generational view stems from the fact that whatever can be written in one language can in principle (given enough ingenuity and time) be written in any other. Lastly, a fourth argument arises from the notion that what is left out of history is often more important than what is officially recorded. Thus, it should be noted that the generational view of computing, with its implicit theme of unitary progress, ignores the conflicts and competition among the different groups within the field and tells us nothing about the theories, techniques, prototype machines or research programmes that have been scrapped. (For a discussion of some of these factors, see Burnett, 1985; Catt, 1973; Fleck, J., this volume; Lamb,

Table 2.1: Alternative schemes for the history of computing

	HARDWARE VIEW	SOFTWARE VIEW	APPLICATIONS VIEW
Generation			
1st	electronic vacuum tubes	stored programs in machine code	numerical calculations for military systems
2nd	transistors	high-level languages FORTRAN, COBOL	general data processing — number crunching
3rd	integrated circuits	structured programming	expert systems in e.g. fault diagnosis
4th	VLSI	shells for knowledge-based systems, software engineering, formal methods	Intelligent Knowledge Based Systems in management
5th	non-Von Neumann architecture parallel processors	symbolic processing with extensive AI applications, AI-based software tools	natural language understanding, inductive and deductive reasoning

1985; Redmond and Smith, 1980; Schopman, this volume.)

While many people in AI and computing regard the imminent 5th Generation as the long-promised embodiment of real artificial intelligence, some go much further and actually envisage the prospect of building machines of super-human intelligence. Among them, Minsky is perhaps one of the more outspoken:

> When and if we choose to build more artfully conceived intelligent machines, we should have many new options that were not available during the brain's evolution, for the biological constraints of vertebrate evolution must have dictated many details of the interconnections of our brains. In the new machines, we will be able to provide whatever paths we wish. Though the new machines still cannot possibly keep track in real time of everything they do, we surely should be able (at least in principle) to make those new, synthetic minds vastly more self-conscious than we are, in the sense of being more profound and having insightful knowledge about their own nature and function (Minsky 1985: 42).

In this, Minsky's belief in the eventual superiority of machines over our human brains (which have supposedly been inadequately endowed with skills by the processes of evolution) is nothing new in the history of computing. It echoes a similar notion held among the System Dynamics Group at MIT. In the latter case the argument was that evolution has not equipped us with the ability to infer how complex nonlinear feedback systems behave but that computers could be used to simulate the behaviour of such systems in a precise and faultless way. From this it was contended that *only* the use of computers could enable us to understand how to formulate policies to cope with the dynamic behaviour of social systems. While the parallel here forces our attention once again on the shared cultural heritage of these two strands of computing, it also illuminates a dream that actually seems to underpin the enterprise of computing in a general sense — namely, the faith that is placed in the ultimate power of technology and the search for a 'forbidden fruit' which always lies 'just ahead'.

2.5 INSTRUMENTAL REASON, HACKERS, AND VISIONARIES

In this section I wish to explore how the picture developed here relates to two of the more influential accounts of AI — namely, those of Turkle and Weizenbaum — both of which offer very different evaluations of AI culture. I will begin with the latter.

Despite his position as a professor of computer science at MIT, as well as being someone who has worked on AI-related research (i.e. language understanding), Weizenbaum has been a frequent critic of the growing cultural fetish with computers which seems to be in the process of coming to dominate Western civilisation. In essence Weizenbaum offers a global theory of the culture of computing — that is, he sees the development of computing in relation to forces within society at large. Thus, he does not view the emerging computer culture as the result of some form of 'impact' between computing and society; instead, the culture of computing is seen as but part of a broader societal drift toward the idolatry of science and technology. Alluding to the writings of the critical theorist Max Horkheimer (1947), once a director and leading exponent of the Frankfurt School, he perceives the current fixation with computers — and hence the obsession with computational calculation rather than human judgement — as evidence of the increasing dominance of instrumental reason within modern societies. Instrumental reason is that form of rationality in which 'means' tend to become 'ends' in themselves; deeply embedded in the development of Western (scientific) culture since the dawn of the Enlightenment, the spread of instrumental reason marks a process in which, among other features, all disciplines seeking a scientific status are driven to seek legitimation by adopting the methods and techniques of the natural sciences. Moreover, within society more generally, the dominance of instrumental reason bodes the increasing rationalisation of social life. Under the regime of instrumental reason, thought becomes reduced to the contemplation of that which can be calculated: 'Thinking objectifies itself to become an automatic, self-activating process; an impersonation of the machine that it produces itself so that ultimately the machine can replace it.' (Adorno and Horkheimer 1972: 25). Written originally in 1944, these words can be viewed as a prophetic anticipation of the genesis of more than one area of contemporary computing.

To support his case that computing is a cultural phenomenon which does not constitute an isolated aberration so much as an exemplar of the present age, Weizenbaum points to other intellectual developments which (for him) also bear the stamp of instrumental reason. He argues that there is a close similarity between the general problem solving approach (GPS) formulated by AI gurus Simon and Newell, Skinnerian behaviouristic psychology, and System Dynamics (the technique I referred to earlier).

As we see so clearly in the various systems under scrutiny, meaning has become entirely transformed into function. Language, hence reason too, has been transformed into nothing more than an instrument for affecting the things and events in the world. Nothing these systems do has any intrinsic significance. There are only goals dictated by tides that cannot be turned back. There are only means-ends analyses for detecting discrepancies between the ways things are, the 'observed condition', and the way the fate that has befallen us tells us we wish them to be. In the process of adapting ourselves to these systems, we, even the admirals among us, have castrated not only ourselves (that is, resigned ourselves to impotence), but our very language as well. For now that language has become merely another tool, all concepts, ideas, images that artists and writers cannot paraphrase into computer-comprehensible language have lost their function and their potency (Weizenbaum 1976: 250).

Almost two decades ago, J.W. Forrester, the father of System Dynamics, asserted that anything which could be stated clearly in words could be incorporated in a computer simulation model; this had the implication that that which could not be represented in a program was in fact unclear. Now, during the current era of rule-based expert systems, we find that human experts (whose expertise is sought for computational representation) are forced to reflect on their own expertise in accordance with the conceptual constraints imposed by the need to represent knowledge as a series of logical rules (Bloomfield, 1986b, c; Collins, this volume; Dreyfus and Dreyfus, 1986).

Weizenbaum's argument enables us to reflect on the nature of computing at a general cultural as well as at a philosophical level (i.e. in terms of science and method); but his argument also has another side, one which requires us to consider the everyday world of computing practice. Doing so provides us with a link between the global view of computing as the embodiment of instrumental reason *par excellence* and the local world of the programmer. In this connection Weizenbaum describes in detail the particular culture of the compulsive programmer — better known as the 'hacker':

> Wherever computer centres have become established . . . bright young men of dishevelled appearance, often with sunken glowing eyes, can be seen sitting at computer consoles, their arms tensed and waiting to fire their fingers, already poised to strike, at the buttons and keys on which their attention seems to be as riveted as a gambler's on the rolling dice (Weizenbaum, 1976: 116).

He argues that compulsive programming is a psychopathological state involving fantasies of omnipotence in which the hacker's sense of power must be constantly verified through the control of the computer: 'He will construct the one grand system in which all other experts will soon write their systems,' (Weizenbaum, 1976: 118). What is of particular interest here is the fact that Weizenbaum sees the culture of hacking as part of a continuum that is represented by the cognitive interest in prediction and control which is common to both science and technology. In other words, there is a link between the imputed pathology of the compulsive programmer and the instrumental reason of science and technology — compulsive programming is but the most extreme form of the interest in prediction and control.

How does Weizenbaum's view of AI and computing square with the argument developed in the preceding sections: is there any scope for synthesis or are the two pictures mutually exclusive? Well, the first comment to make is that Weizenbaum's view presents us with the more global view of AI and computing: despite the continuity just referred to — i.e. between the role of instrumental reason in Western culture and compulsive programmers — his account does not allow for the heterogeneity which is possible among the groups who adapt the ideas and methods of AI and computing. In this regard perhaps the role of instrumental reason and its scientific expression in the form of the cognitive interest in prediction and control are ascribed too large a role. Perhaps the best way to illustrate this is to point out one of the important *differences* between AI and System Dynamics — for despite their common ancestry, it has been shown elsewhere (Bloomfield, 1986a) that in the later stages of its development, System Dynamics veered away from instrumental rationality and took on an increasingly moral stance. When tackling the problems of urban decay and world population growth, the System Dynamicists eschewed scientific or technological solutions and suggested policies that were tied in with changes in value structures. In other words, a group trained among one of the foremost technocratic elites of the modern world opined policies which were the antithesis of instrumental or technological rationality. Thus, the development of System Dynamics cannot be understood in isolation from instrumental reason but neither can it be reduced to it. The lesson here is that AI might also be incorporated into very different philosophies and research programmes; members of its esoteric and exoteric circles might simultaneously be members of other thought collectives. That said, Weizenbaum's argument does help us to understand some of the cultural roots of AI and an important element in the dynamic that lies behind the thought collective which embraces it.

In contrast with Weizenbaum, Turkle offers a rather different view of computer culture. In particular, she lays great emphasis on the distinction between the culture of the hacker and that of the AI scientist: in each case the computer and computing are said to have a different meaning and purpose:

> The hacker culture is isolationist. The computer offers hackers a way to build walls between themselves and a world in which they do not feel comfortable. The culture of artificial intelligence is imperialistic. Here too there are walls that create a sense of being in a place apart. But this time the walls are felt as a fortress from which to conquer the world rather than as a protective shield to keep it at a safe distance (Turkle, 1984: 259-60).

There are several problems associated with this dichotomy: firstly, in Fleckian terms, it ignores the essential intertwining between the knowledge and social bonds of the thought collective of AI. Hence, Turkle neglects to consider the possibility that the conceptual (and indeed social) walls which sustain the 'fortress' mentality or thought style of AI also contribute towards sustaining its knowledge claims; for in fact AI people want to 'conquer the world', but do so only by creating models of the outside within the safe confines of their computer generated micro-worlds inside the 'fortress'. Secondly, it does not allow for the case of those AI scientists who also share some of the predispositions of the hacker. Turkle's interviews with hackers centred on those who were by and large academic drop-outs — that is, people who had curtailed their academic careers in order to devote themselves exclusively to the machine; and when turning to deal with the AI community, Turkle did not pause to ask whether or not some of its members might themselves be really hackers at heart. In fact it might be that there is no one group (or type of individual) which constitutes the real hackers; rather, it seems that the term 'hacker' is often used among people in computing and AI alike to refer to groups or individuals who are seen to be involved on lower-status projects. Whenever one talks to people in the field it seems that the hackers are always 'elsewhere'; this phenomenon may well indicate that the ascription of 'hacker' is an example of projection in which individuals project onto others those traits (e.g. 'loving the machine for itself' (Turkle, 1984: 201-46)) which they would most hate to admit as part of their own make-up.

A third problem with Turkle's dichotomy is that drawing the distinction between hackers and academics has the effect of preserving a privileged epistemological status for AI: hackers have an intense relationship with computing because it satisfies their social and psychological needs; whereas AI scientists are preoccupied with computing because they believe that it offers a new and powerful means of describing (eventually) the entire world. Moreover, given the generally uncritical style of Turkle's book, the reader might be led to draw the stronger conclusion that the AI community is correct in its beliefs, thus justifying its intense involvement with computers. Indeed, such a conclusion is implicit in certain passages where Turkle discusses the differences between AI exponents such as Minsky and opponents such as Searle. Thus, for example, though she opines the idea that the disparities are essentially due to opposing cultures — those for whom the mind is mechanism and those for whom it is not — she again uses divergent forms of explanation to account for the different sets of beliefs. On the one hand, exponents of AI are seen as constructing a new form of world hypothesis, stimulated by the computer's second nature as 'reflective medium and philosophical provocateur'. On the other hand, opponents of AI are seen to be interested in preserving deeply felt commitments about the uniqueness of the human spirit. Her bias on this point becomes even more explicit when she compares the reception meted out to AI with that given to psychoanalysis earlier this century (Turkle, 1984: 328-38).

Even if we went along with Turkle's argument and granted that hackers and academics might have different individual reasons for their intense involvement with computing, the argument presented in the previous sections would indicate that they have much else in common which cannot be simply disregarded when considering their relationship to computers. As exoteric members of the thought collective of AI, many of the hackers who Turkle interviewed share in the thought style along with the esoteric circle. At another level, Turkle's distinction between hackers and AI scientists ignores the fact that a programmer's potential ego-involvement with the programs that he or she creates is a perpetual problem in computing. It is particularly fostered in those programming environments (e.g. academia) where computing is largely an *individual* as opposed to a *social* or collective activity. This point is stressed in Weinberg's study of the psychology of computer programmers: 'Although they are *detached* from people, they are *attached* to their programs. Indeed, their programs often become extensions of themselves' (Weinberg 1971: 53). Programmers' attachment to programs makes them reluctant to admit (especially to themselves) that their code contains errors or 'bugs', with the upshot that there is always a temptation to fix an errant program by local piecemeal or *ad hoc* adjustments (patching) rather than by considering the fundamental revision or even scrapping of a program. To be sure, people working in all areas of science and mathematics are susceptible to ego-involvement with their work — computer programmers are not in any way perverse or unique in this matter. However, work in computing is different in certain key respects which do have a bearing on the problem of attachment. I have already mentioned the relative social isolation of programming, but another important factor is that programmers are often in the situation of manipulating an abstract universe rather than interacting with the real physical one which surrounds us. This point is forcefully put by Weizenbaum:

> (T)he game the computer plays out is regulated by systems of ideas whose range is bounded only by the limitations of the human imagination. The physically determined bounds on the electronic and mechanical events internal to the computer do not matter for that game — any more than it matters how tightly a chess player grips his bishop or how rapidly he moves it over the board (Weizenbaum, 1976: 112).

Moreover, the fact that this state of affairs is positively valued in the esoteric circle is well documented by Turkle's fieldwork:

> For many programmers, much of the excitement of the computer comes from freeing the mind from the constraints of matter . . . What is most valued and most beautiful is what is most freely constructed. In the AI culture that has grown up around Minsky at MIT, the greatest pleasure is building worlds out of pure mind. And his students speak long and lovingly of their earliest experiences of doing this (Turkle, 1984: 261).

In terms of Fleck's model of science, the implication of this state of affairs is that in those areas of AI into which physical reality does not impinge, the active connection imposed by the thought style play a greater role than those passively imposed by nature. The particular problems engendered by this imbalance in active and passive connection can be illustrated by referring again to that earlier era in the annals of computing which dealt with the simulation modelling of social systems. The urban and global models built by Forrester *et al.* at MIT (Forrester, 1969, 1971) were predicated on the philosophy that system structure was more important than data — to the extent that empirical testing and parameter estimation, the benchmarks of modelling in many other areas, were virtually eschewed. Within the artificial world supported by the computer, the System Dynamics models were used to explore various policy alternatives and a number of recommendations were made which, far from being objective or detached, largely reflected the ideological commitments or cosmology of the particular modellers involved (Bloomfield, 1986a). In other words, the abstract simulation models were in effect unconstrained by the resistance of reality. Yet the outputs from the simulations were presented as practicable policies for solving complex social, economic, and political problems. The lesson here is salutary; but the social context within which the thought collective of AI finds itself operates so as to keep at bay the counterexamples to its project which are readily available from disciplines with far greater intellectual pedigrees — disciplines which draw upon decades if not centuries of theoretical and empirical work (Dreyfus, 1981; Shanker, this volume).

Finally, I wish to come back to the argument that AI offers a revolutionary view of the human mind and brain — if not the very nature of the human species — and that by implication its exponents are visionaries. In particular, I wish to examine the notion that the computational model of the mind is more scientific and more detached than more traditional worldviews. To substantiate her case, Turkle develops a comparison between the social impact of AI and those impacts due to the revolutionary heliocentric cosmology of Copernicus, the Darwinian theory of evolution, and the Freudian theory of the mind. She argues that each of these historical cases represented a new system of ideas which challenged existing anthropomorphic sentiments about the centrality of humanity, but that people had always found a way to reassert their sense of importance. Thus, while the Copernican revolution challenged the idea that the Earth and humankind were at the centre of the universe, we could still think of the solar system as being at the centre of things. When Darwin threatened our image of ourselves by demonstrating our kinship with the animals, we could seek solace in the idea that we were the crowning achievement of God's creation. As Freud argued that we were prey to our unconscious urges and desires, human rationality could be portrayed as a triumph of reason over the instinctual; and now, with developments in computing and AI, we have another threat to our sense of self:

Where the Freudian vision seemed speculative to some, literary to others, the computational model arrives with the authoritative voice of science behind it — and with the prospect that someday there will be a thinking machine whose existence will taunt us to say how we are other than it (Turkle, 1984: 322-3).

Turkle contends that unthinkable ideas — threats to our sense of self — become accepted and taken for granted because of a compensatory process by which people re-establish their sense of primacy. However, this view of the human response to developments such as the Copernican revolution appears far too simple: for instance, in contrast to Turkle, Elias (1971a, b) argues that a condition for the emergence of the new heliocentric cosmology was the ability of people to see themselves and their place in the universe in a more detached way: they had to be capable of an act of 'self-distanciation'. The emergence of that ability marked a stage in the evolution of the psychic structures of the human personality and was intimately bound up with long-term processes of social development: people's self image is not a property of individuals so much as a product of society (see Elias, 1978b, 1982a). Turkle, however, presents us with a picture of a rather timeless human personality which always secures the centrality of the self in the face of threatening worldviews. Yet Turkle's account is not just to be challenged because of its a-historical, and indeed a-sociological view of human psychic make-up and the collective (social) representations which have been constructed to explain and give meaning to the world. The more significant problem here is that it begs the question as to whether the computational model of mind actually requires a further act of self-distanciation on the part of humankind. Can we really say that the followers of AI are more detached in their view of the mind than are their opponents: what evidence do we have that this is in fact the case? On philosophical grounds, as argued, for example, by Shanker (this volume), we would have to answer negatively; and in sociological terms also, we find no greater detachment on the part of AI. The argument pursued here — which in fact draws upon Turkle's own fieldwork — reveals how the beliefs of AI are bound up with the imperialism of the whole tradition of AI research, with its roots in the successful application of the formal methods of science and mathematics to technological systems, and with the strong conviction that AI offers the more scientific view of the mind. Moreover, I have argued that the internal and external social bonds of the AI thought collective are mutually reinforcing. On this account one might legitimately turn the earlier question around and ask whether it is members of the AI thought collective who need to self-distance *themselves* from their own particular culture (and detach themselves from their computer programs) in order to comprehend the criticisms of the opposition?

2.6 SUMMARY

In this chapter I have sought to say something about the culture of AI: drawing upon the sociology of science I have presented an account of the AI thought collective and its thought style and discussed the ways in which these are interrelated. I have tried to illuminate the context of judgement within which the assertions of AI — such as the view of the 5th Generation or claims about machine intelligence — make sense or are to be understood. In one way or another, many of the previous accounts of AI which have attempted to address social issues can be considered as integral parts of the thought style; in holding up a mirror to AI they have simply reiterated and reinforced many of its primary beliefs and convictions. In particular, the assumption of technological determinism, coupled with the restriction placed on the epistemological role of sociology (and indeed related disciplines such as anthropology), has merely served to foreclose discussion of the reasons why those in AI believe what they believe. This is the fatal flaw in Turkle's ethnographic study: in effect, her's is an exoteric account of AI, for her argument depends on the interpretations of the subject offered by her interviewees — that is, the esoteric theorists and exoteric hackers who constitute the AI community at MIT.

What has emerged from the alternative picture developed here is the existence of strong internal beliefs and social bonding among the esoteric circle of the thought collective; while at the same time the relationships with other disciplines and traditions are often strained and conflictual. This cultural dynamic — internal cohesion and external dissension — is a mutually reinforcing process. Moreover, the messianic zeal of AI adherents, in particular their conviction in the universality of the concept of a program, coupled with the firm belief that they are leading the way over the threshold toward a new scientific view of the mind and brain, serve to obviate any necessity either to pay attention to, or assimilate, the criticisms of other disciplines. At an institutional level, in terms of academic journals and textbooks, as well as the conference agendas of AI, one finds an orientation toward technical matters; yet, considering the relative newness of AI together with the vastness of its intellectual project, this would seem to be contradictory. With a focus on programs rather than substantive issues, even those who wish to take on board influences from different academic areas can really only adopt those aspects which can be transformed into computationally tractable constructs. If the editorial board of a journal is only willing to accept papers which describe the working of an AI program, then a pre-occupation with technical proficiency can become a condition *sine qua non* for individual academic advancement.

At times it would appear that exponents of AI seem to accept the 'pre-scientific' stature of the subject. Dreyfus, for example, reports that Minsky and Papert have in effect admitted that the relationship of AI to a real scientific understanding of the mind is similar to that which once existed between astrology and astronomy: 'Just as astronomy succeeded astrology, following Kepler's discovery of planetary regularities, the discoveries of these many principles in empirical explorations on intellectual processes in machines should lead to a science eventually' (Minsky and Papert, 1973; quoted in Dreyfus, 1981). Yet we know that the science of astronomy did not evolve as a result of the mere refinement of the then-existing technical instruments. Moreover, we should be aware of the implicit determinism of Minsky and Papert's statement.

The social implications of AI depend upon many factors. Among them there is the extent to which the thought collective succeeds in establishing itself (in struggle with other collectives and disciplines) as *the* legitimate means of orientation in matters pertaining to knowledge and the mind. As the thought collective of AI grows in size it does so both esoterically and exoterically. Therefore, the social implications also depend on the particular popularisations of exoteric groups, which, in turn, are related to the extent to which the esoteric circle conceals the difficulties and failures of AI. Thus, to take but one example, while knowledge engineers are forced to fit expertise into the representational strait-jacket of formal rules and heuristics, the very source of such knowledge, the human experts, become exoteric members to the extent that they come to redefine their knowledge and skills in rule-based-theoretic terms. Yet the severe shortcomings of rule-based representations are already well known among many in the esoteric circle.

Thus, people who fear that one day we will be surrounded by intelligent machines are misplacing their concern; for instead of worrying about the future they might be better employed in considering the current subtle ways in which *we* are becoming more like computers as people adopt (unwittingly or not) the ways of thinking and speaking which are the hallmark of AI.

For many years, game playing, particularly in chess, has been an important test bed for AI research, yet ironically, common-sense knowledge has turned out to be much more intractable. To move from problems of chess to those posed by the need to represent the knowledge and skills which each one of us takes for granted in every daily interaction is to leave the world of the computer laboratory — with its greater freedom for active versus passive connections — and confront the restraints of social reality. The difficulties thereby encountered (Dreyfus, 1981; Collins, this volume) indicate that not only must AI scientists physically leave the computer laboratory behind, but more importantly, their attachment to the culture which sustains it.

NOTES

1. One of the reasons why sociology was excluded from the internal study of science was because of the assumed tension between knowledge on the one hand and society on the other; it was felt that sociology should only enter the picture when knowledge turned out to be false, when it had deviated from the rational path or immanent logic of its development. Modern sociology of science thus gives us a more realistic picture in the sense that it shows us science more as it is rather than as we (or philosophers of science) might wish it to be.

2. What is of particular interest in this account is the competition and conflict between the various groups involved and how these influenced the prevailing conceptualisations of computing machines.

3. An indication of the range of goals and objectives of the 5th Generation projects is given below:

THE RANGE OF 5th GENERATION PROJECTS

(Source: *Computer Design*, 1984; Feigenbaum and McCorduck, 1984).

Country	Organisation	Funding
Japan	ICOT — Institute for New Generation Computer Technology	c. $30 million per yr

(Japanese Government + 8 member
corporations)
Goal: market leadership. Objectives: parallel inference machine,
knowledge-based machine.

Japan   National Superspeed Computer Project c. $20 million per yr
(Japanese Government + other
organisations)
Goal: increase computer speed by 1000 times. Objectives: various.

USA   DARPA — Defense Advanced   c. $1 billion
Research Projects
Agency
(US Government)
Goal: national security. Objectives: speaking pilots assistant,
autonomous intelligent vehicles, battle management systems

USA   MCC — Microelectronics and   c. $50–75 million per
Computer Technology   yr
Corporation
(18 corporations)
Goal: market dominance. Objectives: microelectronics packaging,
CAD/CAM, software development, new architectures.

USA   SRC — Semiconductor Research   c. $15 million per yr
Cooperative
(30 corporations)
Goal: semiconductor industry support. Objectives: numerous.

Britain   ALVEY — Programme for Advanced c. $500 million
Information Technology
(British Government + other
organisations)
Goal: improve competitiveness. Objectives: VLSI, software
engineering, IKBS, man-machine interface.

EEC   ESPRIT — European Strategic   c. $1.5 billion
Programme on
Research in IT
(EEC + other organisations)
Goal: improve competitiveness. Objectives: microelectronics,
software development, information processing, office automation,
CAM.

4. Lilienfeld (1978) sees a connection between AI and the rise of
systems theory; he traces the latter to the development of bureaucratic
capitalism and the goals of administrative and technocratic elites.

REFERENCES

Adorno, T. and Horkheimer, M. (1972) *Dialectic of enlightenment*.
Herder and Herder, New York
Athanasiou, T. (1985) Artificial Intelligence: cleverly disguised poli-

tics. In T. Solomonides and L. Levidow (eds.), *Compulsive technology*, Free Association Books, London, pp. 13–35

Barnes, B. (1982) *T.S. Kuhn and social science*. MacMillan, London
—— and Shapin, S. (eds) (1979) *Natural order*. Sage, Beverly Hills

Bloomfield, B.P. (1986a) *Modelling the world: the social constructions of systems analysts*. Blackwell, Oxford
—— (1986b) Epistemology for knowledge engineers. *Communication and Cognition — Artificial Intelligence, 3(4)*, 305–20
—— (1986c) Capturing expertise by rule induction. *Knowledge Engineering Review, 1(4)*, 30–6

Boden, M. (1984) Impacts of Artificial Intelligence. *Futures, February*, 60–70

Bolter, J. (1984) *Turing's man*. Gerald Duckworth, London

Burnett, P. (1985) Unfair handicaps in the IT stakes: *Times Higher Education Supplement*, 22 March

Catt, I. (1973) *Computer worship*. Pitman, London

Collins, H.M. (1985) *Changing order: replication and induction in scientific practice*. Sage, London
—— (1986) The Turing Test: sociological approaches. Paper presented to the annual conference of the Society for Social Studies of Science, Pittsburgh, 25–28 October

Computer Design (1984), *23(10)*, Sept.

Computer Professionals for Social Responsibility (1984) *Strategic computing — an assessment*. CPSR, Palo Alto

Computing (1986) 26 July, 1

Dreyfus, H.L. (1972) *What computers can't do*. Harper and Row, New York
—— (1981) From micro-worlds to knowledge representation. In J. Haugeland (ed.), *Mind design*, Bradford Books, Montgomery, Vermont; pp. 161–204
—— and Dreyfus, S.E. (1986) *Mind over machine*. Free Press, New York

Durham, T. (1986) Putting intelligence on a more formal footing. *Computing*, 16 October, 28–9

Elias, N. (1971a) Sociology of knowledge: new perspectives, part one. *Sociology, 5*, 149–68
—— (1971b) Sociology of knowledge: new perspectives, part two. *Sociology, 5*, 355–70
—— (1978a) *What is sociology?* Hutchinson University Press, London
—— (1978b) *The civilizing process*. Volume one — *The history of manners*. Basil Blackwell, Oxford
—— (1982a) *The civilizing process*. Volume two — *State formation and civilization*. Basil Blackwell, Oxford
—— (1982b) Scientific establishments. In N. Elias, H. Martins and R. Whitley (eds), *Scientific establishments and hierarchies*, Sociology of Sciences, vol. VI, Reidel, Dordrecht, pp. 3–69

Feigenbaum, E.A. and McCorduck, P. (1984) *The Fifth Generation*. Pan, London

Fleck, J. (1984) Artificial Intelligence and industrial robots: an automatic end for utopian thought? In E. Mendelsohn and H. Nowotny (eds), *Nineteen eighty-four: science between utopia and dystopia* Sociology of the Sciences, vol. VIII, Reidel, Dordrecht, pp. 189–231

Fleck, L. (1979) *Genesis and development of a scientific fact.* University of Chicago Press, Chicago

Forrester, J.W. (1961) *Industrial dynamics.* MIT Press, Cambridge, Mass.

—— (1969) *Urban dynamics.* MIT Press, Cambridge, Mass.

—— (1971) *World dynamics.* MIT Press, Cambridge, Mass.

Foucault, M. (1970) *The order of things.* Tavistock, London

Gardner, H. (1985) *The mind's new science.* Basic Books, New York

Gurstein, M. (1985) Social implications of selected Artificial Intelligence applications. *Futures, December*, 652–71

Hayes-Roth, F. (1983) Using proofs and refutations to learn from experience. In R.S. Michalski, J.G. Carbonell and T.M. Mitchell (eds), *Machine learning*, Tioga Publishing, Palo Alto, pp. 221–40

Hofstadter, D.R. (1978) *Gödel, Escher, Bach: an eternal golden braid.* Basic Books, New York

Horkheimer, M. (1947) *The eclipse of reason.* Oxford University Press, New York

Kling, R. (1980) Social analyses of computing: theoretical perspectives in recent empirical research. *Computing Surveys, 12 (1)*, 61–110

Knorr-Cetina, K.D. and Mulkay, M. (eds) (1983) *Science observed.* Sage, London

Kraft, P. (1979) The industrialization of computer programming: from programming to Software Production. In A. Zimbalist (ed.), *Case studies in the labor process*, Monthly Review Press, New York, pp. 1–17

Kuhn, T.S. (1962) *The structure of scientific revolutions.* University of Chicago Press, Chicago

Lakatos, I. (1976) *Proofs and refutations.* Cambridge University Press, Cambridge

Lamb, J. (1985) Whatever happened to Alvey? *New Scientist*, 21 March, 10–11

Lehnert, W.G. (1985) The PROLOG problem. *Communication and Cognition — Artificial Intelligence, 2(4)*, 3–12

Leith, P. (1986) Fundamental errors in legal logic programming. *Computer Journal, 29 (6)*, 545–54

Lenat, D. (1977) The ubiquity of discovery. *Artificial Intelligence, 9(3)*, 257–85

Lighthill, Sir J. (1973) Artificial Intelligence: a general survey. In *Artificial Intelligence: a paper symposium*, Science Research Council, London, pp. 1–21

Lilienfeld, R. (1978) *The rise of systems theory.* John Wiley, New York

Linn, P. (1985) Microcomputers in education: living and dead labour. In T. Solomonides and L. Levidow (eds), *Compulsive technology*, Free Association Books, London, pp. 58–101

McCorduck, P. (1979) *Machines who think.* W.H. Freeman, San Francisco

Machine Intelligence News (1985) vol. 1(7), April

Mannheim, K. (1936) *Ideology and utopia*. Routledge and Kegan Paul, London

Marcuse, H. (1978) Some social implications of modern technology. in A. Arato and E. Gebhardt (eds), *The essential Frankfurt School reader*, Basil Blackwell, Oxford, pp. 138–62, 180–2

Michie, D. and Johnston, R. (1985) *The creative computer*. Penguin, Harmondsworth

Minsky, M. (1975) A framework for representing knowledge. In P.H. Winston (ed.), *The psychology of computer vision*, McGraw-Hill, New York, pp. 211–80

—— (1979) Computer science and the representation of knowledge. In M.L. Dertouzos and J. Moses (eds), *The computer age: a twenty-year view*. MIT Press, Cambridge, Mass., pp. 392–421

—— (1985) Why people think computers can't. In D.P. Donnelly (ed.), *The computer culture*, Associated University Press, London, pp. 27–43

—— and Papert, S. (1973) *Artificial Intelligence*. Condon Lectures, Oregon State System of Higher Education, Eugene, Oregon

Noble, D.F. (1984) *Forces of production*. Alfred A. Knopf, New York

Pinch, T.J. and Bijker, W.E. (1984) The social construction of facts and artefacts: or how the sociology of science and the sociology of technology might benefit each other. *Social Studies of Science, 14*, 399–41

Pollitzer, E. and Jenkins, J. (1985) Expert knowledge, expert systems and commercial interests. *International Journal of Management Science, 13(5)*, 407–18

Redmond, K.C. and Smith, T.M. (1980) *Project Whirlwind*. Digital Press, Bedford, Mass.

Searle, J. (1980) Minds, brains and programs. *The Behavioral and Brain Sciences, 3*, 417–24

—— (1985) *Minds, brains and machines*. BBC Publications, London

Shallis, M. (1985) *The silicon idol*. Oxford Unidversity Press, Oxford

Shanker, S.G. (1987) The decline and fall of the Mechanist Metaphor. In R. Born (ed.), *The case against AI*, Croom Helm, London, pp. 72–131

Shapin, S. (1982) History of science and its sociological reconstructions. *History of Science, XX*, 157–211

Simon, H. and Newell, A. (1958) Heuristic problem solving: the next advance in operations research. *Operations Research, 6*, 1–10

Snow, C.P. (1959) *The two cultures*. Cambridge University Press, Cambridge

Solomonides, T. and Levidow, L. (eds) (1985) *Compulsive technology*. Free Association Books, London

Suchman, L. (1986) Action in cognitive science. *ISL Technical Note* (Palo Alto Xerox Corp.)

Turkle, S. (1984) *The second self*. Granada, London

Vries, G.H. de and Harbers, H. (1985) Attuning science to culture. In T. Shinn and R. Whitley (eds), *Expository science: forms and functions of popularisation*, Sociology of the sciences, vol. IX,

Reidel, Dordrecht, pp. 103–17

Weinberg, G.M. (1971) *The psychology of computer programming*. Litton Educational Publishers, London

Weizenbaum, J. (1976) *Computer power and human reason*, W.H. Freeman, San Francisco

Woolgar, S. (1985) Why not a sociology of machines? The case of sociology and artificial intelligence. *Sociology, 19(4)*, 557–72

3

Development and Establishment in Artificial Intelligence

James Fleck

3.1 INTRODUCTION

In this chapter, I discuss the role played by scientific establishments in the development of a particular scientific specialty,[1] Artificial Intelligence (AI), a computer-related area which takes as its broad aim the construction of computer programs that model aspects of intelligent behaviour. As with any discussion of a scientific specialty, the identification of what is involved is not unproblematic, and the above serves as an indication rather than a definition. While the term 'Artificial Intelligence' is used in a variety of ways,[2] there is a discernible group (perhaps approaching the degree of commonality to be called a community) of researchers who recognise the term as descriptive of a certain sort of work, and who, if they themselves are not willing to be directly labelled by the term, can locate themselves with respect to it.

Unfortunately, there is little or no commonly available literature that systematically charts the scope of this area. It is worthwhile, therefore, to consider the distinctive socio-cognitive characteristics of research in AI as a prelude to a fairly specific discussion of the social and institutional processes involved in the development of the area,[3] thus providing a basis for exploring the usefulness and applicability of the concept of establishment.

3.2 SOCIO-COGNITIVE CHARACTERISTICS OF ARTIFICIAL INTELLIGENCE

The patterns of research in AI exhibit distinctive characteris-

106

tics, forming a paradigmatic structure which includes such elements in the scientific activity as research tools, practices, techniques, methods, models, and theories, as well as the normative and evaluative aspects for selecting among them.[4] They serve as guidelines and a basis for future research, but are complexly inter-related, often encompassing contradictory facets in tension.

The elements in the paradigmatic structure of AI are as follows:

(1) The general-purpose digital computer provides an instrumental base and a disciplinary context — computer science — for research in the area. Adequate computing facilities are essential for AI work, and hardware limitations have had a constraining effect. Consequently, the availability of funding has been of crucial importance for the development of the area.

(2) List processing languages, a subset of the high-level programming languages available for exploiting the power of the computer, have been developed as tools for research in AI. The community of people using list processing languages, such as LISP in the United States, or POP-2 in Britain, and their variants, can serve as a first approximation for the AI community.

(3) These list processing languages are orientated towards non-numerical uses, and, hence, contrast with more conventional programming languages such as FORTRAN or ALGOL which are numerically oriented. This non-numerical emphasis, with a focus on logic and structure rather than number, distinguishes AI from areas such as pattern recognition, for example, which depends heavily on the use of statistics.

(4) Associated with list processing languages, there has developed a distinctive body of craft knowledge. A high level of skill, gained through first-hand use and practice, is required for the effective use of any programming language and there are many 'tricks of the trade' as well as distinct programming styles, which can only be absorbed through an extended period of apprenticeship.[5]

(5) Embedded in this craft knowledge are numerous elements such as techniques for problem solving, for representing knowledge, for achieving learning ability,

etc. Particularly important and well developed among these are procedures for carrying out searches, often employing rules of various kinds —heuristics — to guide the search and cut down the possibilities to be explored.

(6) The craft knowledge of AI is deployed in the construction of computer program models — computational models — of some aspect of intelligent activity. These models are generally pitched at the symbolic level of meanings rather than at the physiological level of the underlying mechanisms. This distinguishes AI from many other cybernetic approaches, and from much computer simulation work. The focus on intelligent behaviour provides a disciplinary context — psychology — but, due to the great variety of social interpretations and applications of the term 'intelligence', specific goals for research are not thereby dictated. This lends AI a similarity with what can be termed instrument or technique-based specialties, such as X-ray crystallography,[6] which are free to be applied to various goals.

(7) Associated with the wide variety of specific examples of intelligent activity that have been modelled, a clear research area differentiation has emerged since the early 1960s in which subareas have developed their own particular specialist guidelines and techniques, focused on their own more circumscribed concerns. The research areas that could be identified in the early 1960s were game playing, theorem proving, cognitive modelling (an emphasis on models with psychological verisimilitude), natural language, machine vision, and a range of specific applications,[7] some of which have themselves subsequently differentiated out into well defined research areas. These research areas (or strands of research[8]) constitute a primary setting for scientific activity, and consequently have been one of the basic arenas for competition among practitioners, as will become evident.

These cognitive characteristics, or elements of the AI paradigm structure are, of course, at a very general level. They open up a huge cognitive space and offers wide opportunities for exploration, which were elaborated at a fairly early stage in essentially their complete form, while subsequent work has largely exploited the possibilities opened up. This overview of the development of AI invites comparison with Edge and Mulkay's account of the development of radio astronomy: the initial discovery of radio waves from space opened up the possibility of a new source of astronomical information — a new cognitive space — which was subsequently exploited by ever more sophisticated methods of detection, leading ultimately to a revolution in the conception of astronomy.[9] However, while radio astronomy was apparently allowed to develop without much external conflict,[10] the same cannot be said of AI.

3.3 COMPETITION AND ESTABLISHMENT IN ARTIFICIAL INTELLIGENCE

Conflict in AI has been bound up with the focus on intelligence. Intelligence is not a socially or cognitively well-defined goal and every distinctive social group tends to have its own implicit definition, couched in terms of its own interests. Consequently research in AI has been oriented towards a variety of goals. This multigoal character leads to a range of struggles between various groups and establishments within and around AI, and is institutionally manifested in the high degree of research-area differentiation, with interdisciplinary and multidisciplinary affiliations, and associated multiple funding sources. This leads to competition on the one hand, between research areas for resources, and on the other hand over the definition of what AI is. This has had quite clear effects on development, as, for example, with the debate in the UK Science Research Council (SRC) in the early 1970s, which led to separate funding mechanisms being set up for cognitive science (linguistics, philosophy, and psychology) applications of AI, and for research within computer science.

This multigoal characteristic, involving competing groups with different aims, has exerted centrifugal pressures on research in the area and has resulted in the non-emergence of a specialty-wide general theoretical dynamic. Attempts at the elaboration of theories of intelligence have informed work in the area — for example, the early programme of research (evident in work in the 1960s in systems such as GPS (General Problem Solver)), towards forming general mechanisms of inference that would embody the essence of intelligence — but these attempts foundered upon the diversity of concepts and applications involved. Later developments such as the attack upon the problem of knowledge representation (a major theme of research in the 1970s) effectively accepted the contingent diversity of intelligence and turned it into a virtue. What theoretical developments there have been, however, have been very specific and localised, often pertaining to the status of the methods and languages employed.

Nevertheless, the absence of a uniform goal or a general theoretical dynamic raises the question of the source of cohesion and co-ordination for research in the area. The answer seems to lie in the craft nature of the paradigmatic structure. While there are many divisions over short- and long-term goals, and between different research areas, there *is* a shared body of technique and practice based on the use of list processing programming languages and transmitted by apprenticeship and personnel migration, thus constraining the historical development of the area. Access to this body of knowledge and skill is restricted by the need for first-hand contact and for adequate computing facilities, consequently leading to tight intercentre and intergenerational linkages in the area. The group of people who control access to these resources clearly constitutes the establishment in AI, and it is at this level that much of the internal research-area competition takes place. It would seem, therefore, that this case demonstrates that scientific establishments need not be characterised by a high degree of solidarity. It would also seem to be the case that a common basis in technique is adequate to hold an area together in the face of strong centrifugal tendencies, especially where it is associated with an instrumental basis for research in the area. The need for adequate computing facilities has restricted access to the field and encouraged the development of a strong communication infrastructure, particularly in the United States where the ARPA (Advanced Research Projects Agency) computer network enables researchers at geographically distant sites to communicate as easily as if they were at the same location. Such expensive instrumental needs have, of course, opened the area up to influence from funding agencies. Research in AI has without doubt depended upon substantial support from various agencies, such as the US Department of Defense, and the UK Science Research Council, and has consequently been shaped by the concentration policies of these agencies which have had the effect of consolidating the position of the establishment in the area. However, it is difficult to find evidence for any positive direction of research by the funding agencies during the 1960s, although in the stringent financial climate of the 1970s this did change, and tighter demands for the attainment of particular goals were made, resulting in a restructuring of funding patterns among AI centres, both in America and in Britain. Moreover, during the early 1980s, there appears an initial emergence of 'AI-technology', where specific lines of research, considered to have commercial potential (not necessarily those of prime scientific importance), are being picked up out of the university context, along with supporting personnel, and transferred for industrial development. At this level of a broad overview of AI, there appear to be some similarities to Yoxen's description of molecular biology in terms such as 'directed autonomy'.[11] However, at the more detailed level of the following discussion, it is very hard indeed to identify elements of long-term direction that might fit in with concomitant long-term strategies on the part of the funding agencies. Perhaps what is at issue here is the appropriate size and nature of the envelope within which autonomy is exercised by the practitioners, while yet remaining suitably circumscribed in accordance with the externally imposed direction.

The availability of funding and institutional resources for the area as a whole is, of course, controlled by a wider establishment — the funding agencies and the universities — and in the following discussion the processes of negotiation between the specialty and wider establishments stand out clearly. It will become clear that the response of the wider establishment of AI is by no means uniform, thus illustrating that too monolithic a character should not be imputed to the establishment at this level either, but that an adequate understanding needs to take into account the particular features of the area — that is, the specificity of AI.

The allocation of resources and the negotiation between the establishments in and around AI have been clearly affected by what Elias termed a struggle for monopolisation of the means of orientation.[12] Research in the area is often seen as constituting a further thrust of mechanical materialist science into an area — the nature of mind — hitherto under the exclusive sway of traditional cultural values. By and large, mind or intelligence is regarded as the most characteristic and unique of human attributes — extremely rich and complex, and undoubtedly

111

beyond the reach of scientific analyses more appropriate for the understanding of inert matter. Thus, the focus on intelligence and mind brings AI into an arena of conflict at a deep-seated emotional level which touches immediately upon everyone's image of themselves, and induces strong 'for and against' alignments. It is an issue of general public rather than narrowly scientific interest, as is evidenced by the high relative exposure AI receives in the press, on television, etc. It is doubtful whether many other scientific areas could have the same effect — with the exception of some areas such as genetic manipulation which undoubtedly bear comparison. The nature of mind is an area where the religious and philosophical establishments still claim authority, and AI has to fight for legitimacy. Even where explicitly religious commitments do not seem to be involved, those brought up under the Western humanist culture often feel threatened by what they see as the reductionist nature of AI, and work in the area has been denounced as bad science, non-science, gross reductionism, and even immoral science. At this level, then, there is negotiation and conflict with establishments outside science, as well as between establishments within science.

As far as competition between establishments within science is concerned, the case of AI illustrates an important, characteristically twentieth-century development in scientific thinking — the software sciences — with a focus on pattern and organisation rather than on the properties of substance or matter. Yoxen points out in his paper the importance of such metaphors as code, information, read-out, program, etc., in the reconstitution of biology: with AI such ideas are at the very core of the subject. Moreover, with the increasing penetration of the computer into all areas of science and scholarship, the features of AI related to the software science nature of research in the area, may well become typical of many fields of science. In particular, the diffuse, method-based character of AI, with its contingent adaptation to diverse substantive issues, poses a contrast and challenge to the coherent, theoretically centred nature of the current scientific ideal, deriving from the example of the dominant physical sciences tradition. The former would not seem to support so readily a monolithic unified establishment as does the latter, and the former may consequently have implications for the future development of the sciences.

Thus, AI appears to be an interesting case in the context of a discussion of scientific establishments, for a number of reasons. The more diversified nature of establishments in the software sciences may have wider implications for the sciences as a whole; negotiation and conflict between establishments at a variety of levels is clearly illustrated — including a struggle for the monopolisation of the means of orientation; the power bases of these various establishments (control over cognitive, instrumental or financial resources) are clearly evident, and finally AI provides many examples of the problems arising from struggles between the various establishments involved in a multi-disciplinary and interdisciplinary area of research — illustrating many of the points commented upon by Elias.

3.4 EARLY DEVELOPMENT IN THE UNITED STATES

The Second World War acted as a melting pot for various quite different lines of research and disciplines. In the intense concentration on the common goal of winning the war, traditional disciplinary boundaries were broached and new areas of research emerged, such as information theory, operations research, cybernetics and, of course, the development of the digital computer itself. These areas of research can be broadly characterised as the software sciences, in that they focused on pattern and organisation rather than on substance or matter — the concern of the natural sciences such as physics and chemistry.

Cybernetics, a rather general field given a name and identity by Norbert Wiener's classic book, *Cybernetics — control and communication in the animal and the machine*, was concerned with the essential similarities between machines and biological processes.[13] Work in the area developed during the 1940s, and involved such approaches as the comparison of biological and neurophysiological processes with electrical circuits and networks of artificial neurons, or the investigation of the general principles of adaptation in self-organising systems — systems which were rich in feedback connections, and would settle into stable configurations after being disturbed.[14]

The advent of the digital computer in the early 1950s heralded a new approach which sought to build models of intelligent processes at the symbolic level.[15] Concepts were represented and operated upon directly in the computer using high-level programming languages. These 'symbolic' models represented intelligent activity at the level of thought itself, rather than at the level of the physiological mechanisms underlying thought, thus contrasting sharply with other cybernetic approaches.

In 1952, a conference was held under the rubric 'Automata Studies'.[16] This conference, organised largely by John McCarthy, was intended by him to attract proponents of the symbolic modelling approach. It failed in this aim, and attracted contributions more clearly in the other cybernetic traditions. This determined McCarthy to 'nail the flag to the mast the next time', which he did by explicitly using the term 'artificial intelligence' in a subsequent summer school held at Dartmouth College in the United States in 1956, to discuss 'the possibility of constructing genuinely intelligent machines'.[17] The official title was 'The Dartmouth Summer Research Project on Artificial Intelligence' and it did succeed in isolating the symbolic modelling theme. Among those present were J. McCarthy, M.L. Minsky, H.A. Simon and A. Newell.[18] After the meeting, Simon and Newell were to start a group at the Carnegie Institute of Technology (now the Carnegie-Mellon University) with the aim of developing models of human behaviour, while McCarthy and Minsky built up a group at the Massachusetts Institute of Technology (MIT), with the goal of making machines intelligent without particular reference to human behaviour. Later in 1962, McCarthy was to move to Stanford University, where he initiated another AI project. These three centres, along with Stanford Research Institute, dominated AI research in the United States in the 1960s and 1970s. Also present were C.E. Shannon (known in AI for his outline of the chess-playing paradigm which is essentially the same as the one underlying the microelectronic machines that can now be bought off the shelf) and A.L. Samuel (who developed an impressive checkers playing program which incorporated an elementary learning mechanism).

It was at that meeting that the broad outlines of a distinctively AI approach — indeed, what might be called a proto-paradigmatic structure — emerged. This involved the use of high-level programming languages to provide symbolic models of various aspects of intelligent activity. The first areas attacked, chosen partly because they seemed to epitomise intelligence, and partly because they were sufficiently well defined to be readily programmable, were theorem proving in mathematical logic, and games such as chess and checkers. While chess and other games employed numerically based techniques for choosing board moves, the 'Logic Theorist' of Simon and Newell (which was presented at the Dartmouth conference, the first working, characteristically AI program developed) employed non-numerical techniques.[19] During the late 1950s, programming languages designed specifically for non-numerical symbolic information processing were developed by those two researchers, along with J.C. Shaw,[20] and in 1960, McCarthy formulated LISP (list processing language) which became, and still is, the most widely used AI language.[21] During the late 1950s, it also became clear that organisational techniques of search were of paramount importance in attaining the desired ends, and that the numerical aspects were of secondary importance. The principle of looking for and using certain heuristics — that is, rules of thumb which might help in finding a solution but which would not guarantee a solution — became established.[22] By the early 1960s, various successful programs had been written, resulting in a general air of optimism, and indeed by this time the paradigmatic structure of AI had been elaborated in essentially its complete form, as already described.

3.5 THE ESTABLISHMENT IN THE UNITED STATES

At first sight it might seem remarkable that McCarthy, Minsky, Simon and Newell, without doubt the four 'great men' of the AI establishment, should all have been present at the Dartmouth meeting. I would argue, however, that the emergence of the American establishment in AI was part and parcel of the process of defining the paradigmatic structure of research in the area and the organisational structure of the field.

In the first place the four were actively involved in the organisation of the field. McCarthy, as already noted, arranged meetings to bring together those interested in the very loosely defined goal of constructing genuinely intelligent machines. Subsequently, he went on to found a group at MIT, along with Minsky (a fellow student with him at Princeton), and later the group at Stanford. Simon and Newell developed the group at Carnegie.

In the second place, these four men were centrally involved

in defining the substantive cognitive elements of the AI paradigmatic structure already outlined. Simon and Newell, as well as presenting the first working AI program, the Logic Theorist, had also developed a series of list processing languages, the IPL series, the forerunners of the basic element of the AI research activity. McCarthy had produced the definitive AI programming language, LISP, on the basis of these forerunners, and had also incorporated certain features which embodied the emphasis on logic rather than numerical mathematics. Minsky had written an influential systematising paper which explicitly outlined the importance of heuristic search. As well as producing the basic tools for subsequent research, these four also defined in broad terms higher-level guidelines for future research. Simon and Newell pioneered the focus on investigating human cognitive processes as a source of inspiration for computational models, while McCarthy and Minsky went in more for the idea of investigating mechanisms for achieving intelligent activity in the abstract, without prejudice towards specifically human forms.

The fact that the four were involved in the formation of AI research activity from the beginning has contributed to their success in becoming members of the establishment in two ways. In the first place, the founders of any field, simply because at that stage they are competing with fewer people, gain a visibility which later contributors are unlikely to attain, unless they in turn can produce work that will lead to subsequent distinctive and fruitful development. In the second place, all four have been around for a long time, and consequently the process of normal scientific career progression has ensured their continued visibility and position within the establishment.[23] Moreover, once recognised as members of the establishment, they have in fact continued to be influential within the field. McCarthy's suggestions for a mathematical theory of computation,[24] and his emphasis on the use of the predicate calculus, have been themes which have been taken up and developed. Minsky has continued to produce influential synthesising research programmes, as for example in his presentation of the theme of semantic information processing,[25] or more recently with his explication of 'Frames' — high level data structures for organising and mobilising the vast knowledge bases with which effective AI programs have had to work.[26] He has also used his establishment position to argue effectively against the formal theorem proving strand of research.[27] Simon and Newell have continued to pioneer new approaches at the psychological interface of AI and the 'production systems' formulation developed out of their earlier work on general problem-solving, and promoted by Newell, has been widely taken up.[28]

However, the members of the AI establishment did not arrive from nowhere. Their success, without doubt, owed much to their having attended prestigious institutions as students, and to their being sponsored by people who were already members of the wider scientific establishment, not necessarily in the cybernetic area. McCarthy, Minsky, and Newell all attended Princeton as graduate students, for instance, and McCarthy worked for Shannon on the organisation of the 1952 Automata Studies conference; while Minsky was associated with W. McCulloch, whose 1943 paper with Pitts is recognised as another of the texts marking the emergence of cybernetics.[29] Simon had already established his reputation in the fields of political science and economics,[30] and he himself acted as sponsor for Newell.

Moreover, this intergenerational establishment reproduction process continued, and students of McCarthy, Minsky, Simon, and Newell have largely dominated the field. Figure 3.1 gives a graphic illustration of the prevalence of such links — links which have led to charges of nepotism being levelled at the AI establishment.[31] This structure of very strong intergenerational linkages turns out to be characteristic of the development in Britain as well, and in the section on the establishment in the United Kingdom, some underlying reasons for the strong linkages are discussed.

The emergence of this group as the establishment in AI was undoubtedly consolidated by its success in getting the backing of the US Department of Defense, mainly through the Advanced Research Projects Agency (ARPA), which provided some 75% of US AI funding for the ten years from 1964, and through the Air Force.[32] Furthermore, the preference on the part of the ARPA for concentrating resources in a few selected centres guaranteed the position of the establishment, especially in view of the great expense of adequate computing facilities, which effectively barred other groups from competing.

Another aspect of development in AI which has been characteristic and of importance for the field, and which has served further to reinforce the position of the establishment,

117

Figure 3.1: The establishment in the United States, 1960—mid-1970s. The members of the establishment were derived from a consideration of the editorial board of the *Artificial Intelligence Journal,* conference organising committees, invited conference speakers and panel members, supplemented by well known researchers as judged on the basis of a reading of the literature. They include 73 out of a total of upwards of 500 contributors to the area. Available data were limited, but indicated that only some 11 out of the 73 had *not* worked at some time or done a PhD at one of the big four AI centres: Massachussetts Institute of Technology (MIT); Carnegie Mellon University (CMU); Stanford University (SU); and Stanford Research Institute (SRI). At least 24 of the 73 received their doctorates from one of MIT, CMU, or SU. There is no particular significance in the ordering, nor is the record of movements complete. Intergenerational and intercentre linkages are probably underestimated due to lack of data.

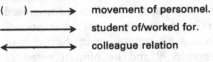

( )——————→   movement of personnel.

——————→   student of/worked for.

←——————→   colleague relation

═══════════   boundary between the wider establishment, and the Artificial Intelligence establishment.

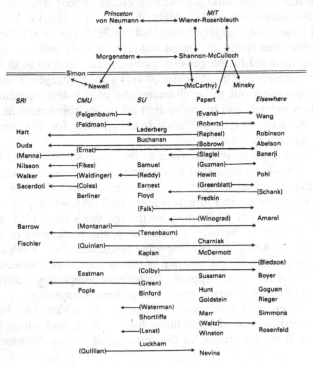

should be noted: this is that the general aim of research in the area to produce intelligent machines has excited extreme reactions and has tended to lead to very strong for-and-against[33] alignments. Such a reaction is not at all surprising given the sensitivity of such a goal to peoples' images of themselves. Here Elias's comments about the competition for the monopoly over the means of orientation are relevant.[34] The AI approach is seeking to establish and legitimate a view of intelligence and the nature of mind which challenges the received commonsense view of mind and intelligence as something rather special and certainly well beyond the reach of scientific analysis. Moreover, this received view is very much under the sway of the religious establishments, or where religious authority does not hold, under the sway of a liberal humanist tradition. Strong reactions are commonplace in AI and, on the sociological level, have probably had the effect of heightening the difference between those on the inside and those on the outside, consequently reinforcing and concentrating the position of the establishment.

Thus, it can be observed that the emergence of the American establishment was very much bound up with the development of AI as a distinctive area of research, and its position was consolidated by the success in gaining backing from the Department of Defense. The American establishment was not only involved in providing an organisational basis for research in the area, but was also very closely concerned with the elaboration of a distinctive cognitive basis for research in the area, the AI paradigmatic structure. In the following discussion of the development of AI in the United Kingdom, some of the themes already introduced will be reiterated, while other issues will become evident.

3.6 DEVELOPMENT IN THE UNITED KINGDOM

In Britain during the 1940s, there was a similar flourishing of interest in general cybernetic concerns as occurred in the United States, and discussions of the possibility of machine thought were common.[35] A.M. Turing was an enthusiast for the possibility of intelligent machines, and his 1947 and 1950 papers still stand in many respects as definitive surveys of the arguments for and against AI.[36] R.J.W. Craik, whose 1943 book *The nature of explanation* is recognised as one of the texts marking the emergence of cybernetics, wrote passages that bore

119

a remarkable foreshadowing of the actual AI paradigmatic structure, as for example in the following passage:

> . . .thought models, or parallels, reality — that its essential feature is not 'the mind', 'the self', 'sense data', nor propositions but symbolism and that this symbolism is largely of the same kind as that which is familiar to us in mechanical devices which aid thought and calculation.[37]

Others were interested in more specific cybernetic approaches: W.R. Ashby, whose name is perhaps second only to N. Wiener's in association with cybernetics, contributed many ideas and books on the subject;[38] W.G. Walter, who achieved a degree of fame with his electronic tortoises which exhibited elementary reflexive behaviour;[39] F. George, who wrote on a cybernetic approach to the brain;[40] and A.M. Uttley, D. Mackay and others, who worked on neural net models of cognition and perception. Furthermore, through the informal RATIO club which existed during the early 1950s (to which Turing, Mackay, Ashby, Walker, Uttley and others belonged) there was frequent interaction and discussion on these issues — discussion which often also involved researchers from the United States: McCulloch, for example, attended the first meeting of the club in 1949.[41]

These discussions continued through the 1950s. There were several British contributors to the 1952 Shannon and McCarthy Automata Studies meeting, and in 1958 a conference on the 'Mechanisation of Thought Processes' was held at the National Physical Laboratory in England, to which people such as McCarthy and Minsky contributed, as well as the proponents of other cybernetic approaches and practitioners of the art of using the digital computer.[42] Indeed, it was there that McCarthy presented his 'Advice Taker', a suggestion for using the predicate calculus for modelling commonsense reasoning, an approach which proved influential within the more specific AI context.[43]

Nevertheless, despite all the activity and discussion in the area, the specifically AI approach was not developed in Britain until the mid-1960s, and then the prime mover was someone quite external to the cybernetic network: this was Donald Michie, a geneticist at Edinburgh University. During the War he had worked with Turing at Bletchley and became fascinated in the possibility of constructing machines that could think.[44] There were no opportunities for him to follow up these interests after the War, except on the hobbyist level, so he took medical sciences at Oxford and subsequently specialised in genetics.[45] In 1962, however, during a visit to the United States (arising out of his hobbyist work on trial-and-error learning in the game of noughts and crosses) he became aware of the developments in AI there and was impressed by the computer facilities available. On his return from the United States, depressed by the lack of such facilities in Britain, he became active in agitating for something to be done — lobbying, writing newspaper articles, and so on.[46] At about the same time, he had grown dissatisfied with his position in the Department of Surgical Science at Edinburgh, where he held the post of reader, and in 1963, very much on his own initiative, he moved out of the department and with his secretary and part-time helpers set up an unofficial unit — the Experimental Programming Unit. Meanwhile, he continued his lobbying for improved computer facilities and had buttonholed C. Jolliffe, the deputy grants director of the then Department of Scientific and Industrial Research (DSIR), whom he impressed with his concern over the state of UK computer provision, and was consequently commissioned to carry out a survey of computing interests and views among British scientists.[47] Lord Halsbury had just been charged with setting up a specialised computer board in the DSIR (subsequently the SRC) with the aim of reviewing and supporting research in computer science. The results of Michie's report indicated a widespread positive assessment of the potential of AI among young computer scientists and this undoubtedly formed the basis for the proportionately generous funding by the computing board of the SRC for research in the area during the mid- to late 1960s.[48] This took place against the background provided by the Flowers Report of 1966 on computing in universities and colleges, which recommended a large expansion in computer provision and training, and set the outlines of the presently existing regional computing centre structure, based in London, Edinburgh, and Manchester.[49] In this context, Michie, with his energy and enthusiasm, was seen very much as a bright young man, and he succeeded in attracting several large grants,[50] which no doubt encouraged the university authorities to give official recognition in January 1965 to his irregularly established unit, which grew rapidly over the next

121

few years.

It is interesting at this point to consider the substantive lines of research that emerged at Edinburgh,[51] for the research profile bore a remarkable similarity to the patterns evident in the United States. Michie's own immediate concerns were with game-playing and heuristic search, then regarded as central to the field. Also, in the Experimental Programming Unit there was a project involved with human problem-solving studies, similar in flavour to the work of Simon and Newell at Carnegie; and there were also projects concerned with developing the instrumental base for AI research. One element was the Mini-Mac project, an interactive multi-access system (and the second such system to be developed in Britain), so called after the first project of its kind, carried out on a grander scale — project MAC at MIT, and also partly motivated by AI concerns, according to McCarthy.[52] Another element in the instrumental base was the development of a list processing language, POP-2, which became the staple AI language in the United Kingdom (and with refinements is still in use today),[53] just as LISP was the staple in the United States.

An independent group worked in the Metamathematics Unit with a focus on automatic theorem proving. This unit had been set up by Bernard Meltzer, a reader in the Department of Electrical Engineering, who had become interested in the use of the computer in the course of his work and combined this interest with his hobby of mathematical logic. Like Michie, he took the fairly dramatic step of moving out of his official department and into a new, unofficial unit after a visit to the United States, where he had visited similar projects.[54] Due to their common use of symbolic rather than numerical information processing, and the intelligent nature of their goal, theorem proving, these approaches had developed as an autonomous but central strand in the AI area of research.

Thus, by the mid-1960s, there had emerged in Edinburgh a centre for research in AI which reproduced the same features as had developed in the United States. In September 1965, Michie organised the first of a series of meetings — the Machine Intelligence Workshops — which were held in Edinburgh and attracted leading AI researchers from the United States, as well as interested people from elsewhere in Britain.[55] These workshops played an important part as a forum for discussion and the communication of the AI approach, and influenced such people as E.W. Elcock and J.M. Foster of Aberdeen University, where in an SRC-sponsored computer unit they worked on game-playing programs and high-level programming languages and systems which incorporated some AI elements.[56] M.B. Clowes was another interested researcher who attended. Encouraged by Michie, he had set up AISB (the Society for Artificial Intelligence and the Simulation of Behaviour) in 1964, which, after an informal and hesitant start has grown into a thriving learned society for the area, holding two yearly conferences and publishing a regular newsletter.[57] Clowes had also visited the United States and had been impressed by what was happening there. He did not work on specifically AI projects in the mid-1960s, and depressed by the lack of suitable computer facilities, he went to Australia for several years. However, he had met and impressed N.S. Sutherland, who had been interested in mechanistic models since the 1950s, and who built up a centre in experimental psychology at the newly established University of Sussex at Brighton in the 1960s. On his return from Australia, Clowes went there to work in AI on a grant held by Sutherland. This work heralded the emergence of Sussex as a major centre for AI in the 1970s.

It was during meetings of the AI community following the technical business of the workshops when suggestions were first put forward for starting a speciality journal for the area — *Artificial Intelligence — An International Journal*, which was eventually founded in 1970 and of which Meltzer became the editor.[58] It also appears that suggestions for establishing international conferences on AI were discussed at Machine Intelligence Workshops:[59] the first was held in 1969, and since then these conferences (held every second year) have grown steadily in size and importance.[60]

Thus, it is clear that the Machine Intelligence Workshops were of great importance for the social development of the field at the international level and consequently, firmly establishing Edinburgh as an AI centre of international reputation. This reputation was further enhanced when Michie succeeded in attracting to Edinburgh from Cambridge the research group of Richard Gregory, the psychologist who became known for his book on the eye and brain,[61] and H.C. Longuet-Higgins, a theoretical chemist of international standing. The basis for the merger was the goal of building an intelligent robot — a goal that was also being pursued at other major research centres in

AI in the United States: MIT, Stanford University, and Stanford Research Institute.

The robot project was seen to pose a challenge for AI in that it required the integration of many of the strands of work within the area — machine vision, problem solving (often based on theorem proving methods), manipulation of a hand in three-dimensional space, and even natural language for communication. Gregory's group was to provide the perception and engineering aspects, while Michie and Longuet-Higgins worked on the problem solving and cognitive aspects. For this project the SRC awarded a major grant and provided a new computer. In addition the Nuffield Foundation provided funding for equipping an engineering laboratory to build the robot and associated hardware and other devices.[62]

Part of the deal involved in attracting these research groups was that the University of Edinburgh, largely as a result of the good offices of Michael Swann (now Lord Swann), then the Vice-Chancellor, would set up a new department — the Department of Machine Intelligence and Perception — and provide chairs for the new senior people involved.[63] This was the first specifically AI-focused department anywhere in the world, and was seen by some as an exciting venture, though others were less welcoming. In particular, as an independent institutional entity in the university, it was in direct competition with the new computer science department, resulting in rather distant and at times antagonistic relations between the two departments.[64] This contrasted with the situation in America, where AI was usually carried on within the departments of computer science and electrical engineering.

This institutional innovation would not have been possible had it not been for two favourable factors. The first was the general context of university expansion of the early and mid-1960s. This expansion, in fact, started being curtailed just after the establishment of the new department, and the resulting squeeze contributed to the problems which beset the department, as the level of University Grants Committee support could not keep pace with the large Research Council funding that the new, rapidly growing area attracted. Had the curtailment of expansion come a few years earlier, it is highly unlikely that a separate department would have been approved. The second factor was the support afforded by Swann: as Dean of Science he had backed Michie's previous initiatives, and newly elected in January 1966 as Principal of Edinburgh University, he was very receptive towards new departures and constantly promoted the status of Edinburgh as second only to Cambridge in research.[65] Without such sponsorship, it is again doubtful whether a new department would have been instituted, or whether Michie would have succeeded in attracting Gregory and Longuet-Higgins. However, the department was established in October 1966, and while in the event the institutional attractions were evident, it is interesting to consider the scientific motivations for these people with a non-AI background to change their area of research.

Gregory had engineering interests which led him to seek a new methodology involving a closer study of the physical basis in the brain for perception and cognition than was usual in psychology at that time, and the AI approach seemed to promise developments along these lines.[66] He also brought with him other members of his group, notably S.H. Salter, who had extensive engineering competence and built the robot hardware (and was later to become known for his wave-power system — the Salter duck); and J.A.M. Howe, a psychologist who was to explore the applications of AI in educational research, and who became head of the AI department at Edinburgh in the late 1970s.

Longuet-Higgins was very much a scientific high flyer, achieving international distinction in his work in theoretical chemistry with C.A. Coulson at Oxford, and gaining a professorship at the early age of thirty. For his eminence in chemistry, in 1958 he was elected a Fellow of the Royal Society, and in 1968 a Foreign Associate of the US National Academy of Science, the highest American honour available to a non-US citizen. Despite his great success in chemistry, or perhaps because of it (in that he was motivated to seek similar success in a new and potentially exciting but unexplored field), he had joined with Gregory in planning a Brain Research Institute. Negotiations were in hand for funding from the Nuffield Foundation, and for accommodation at Sussex University, when Michie, at a meeting with Gregory early in 1966, suggested that Edinburgh would be an ideal centre in view of its already established AI work. With the institution of the new department at Edinburgh, Gregory's group and Longuet-Higgins moved to Scotland, and great hopes were entertained for the future of co-ordinated research in the area.

The anticipated co-operation failed to materialise. Problems over accommodation, personal, political, administrative, and scientific factors were involved in what became a very complex and confused situation during the late 1960s and early 1970s. Gregory never really settled in at Edinburgh nor became involved with the computational approach, though he remained favourably inclined towards it, and in 1970 he finally left to go to Bristol University. The engineering workshop in the Bionics Laboratory had proceeded, however, with the building of the robot hardware, and a prototype was connected to the computer for the first time in May 1969. Longuet-Higgins did absorb the computational approach, but irreconcilable differences between him and Michie over the installation of the new computer and the robot project, as well as over their approaches to work in the area, soon emerged and resulted in Longuet-Higgins moving into separate accommodation and thenceforth running his unit (then called the Theoretical Section) quite independently apart from access to the common facilities.

Michie favoured a rather swashbuckling style of directing large team projects oriented to goals which could be linked with industrial applications and, in fact, was involved in launching a university-based company to market compiler systems and other software for the POP-2 language, which was developed in the department.[67] In addition, he was extremely energetic and persuasive and very successful in obtaining funding from many different sources.[68] Longuet-Higgins, in contrast, favoured a more restrained, academic style, preferred an individual basis of working with a few colleagues on research chosen purely for its intrinsic scientific interest, and was dubious about the advisability of mixing commerce and industry with research. These differences in style, aggravated by contrasts in personality, were associated with conflicting views on AI: Longuet-Higgins thought that 'artificial intelligence' was not a science or technology in its own right, but was a new way of tackling problems in those existing sciences which were relevant to the phenomena of intelligence. It set new standards of precision and detail in the formulation of models of cognitive processes, those models being open to direct and immediate test.[69] Michie's position, on the one hand, was closer to the view

. . .that success in achieving the long-term aims of Machine

Intelligence should be regarded as the major goal of Computing Science. Furthermore, progress in Machine Intelligence is continually generating pressures for solutions to fundamental problems of Computing Science in an environment where they will be used; an environment which by its very nature, demands quality and generality. As a result a decision to invest apparently disproportionate sums into Machine Intelligence could only have beneficial effects to the whole of Computing Science.[70]

This lack of consensus among the practitioners within AI was undoubtedly a complicating factor when the SRC Computing Science Committee came to review its funding policy in the early 1970s. The 'cognitive science' view of AI put forward by Longuet-Higgins was not seen by the reviewing panel as falling within its scope, while there was strong opposition to the view that machine intelligence should be regarded as a major goal of computing science. These differences in views were noted in the SRC *Computing Science Review* of 1972, and were given by Sir Brian Flowers, then chairman of the SRC, as being among the reasons for commissioning Sir James Lighthill to review the field,[71] a review that was to have a large impact on the area as will be discussed in due course.

The differences between Michie and Longuet-Higgins also caused great internal problems at Edinburgh, resulting in frequent appeals being made to the Principal and Secretary of the university, and to the SRC, and a bewildering sequence of organisational forms were instituted by the university authorities in attempts to alleviate the embarrassing situation; but to no avail. Michie, however, started losing the support of his other colleagues, and finally, in 1974, following a lengthy and, to Michie, unsatisfactory, review of the situation,[72] a new Department of AI was set up with Meltzer at its head and comprising most of the AI research groups in Edinburgh. Michie was given his own independent Machine Intelligence Research Unit. The bulk of the very considerable resources which had been built up over the decade, including the robotic equipment, was settled with the department, and, furthermore, limits were placed upon the scope of future research efforts by Michie.

However, despite the tension between the senior people at Edinburgh, there was a thriving research environment in the late 1960s and early 1970s, with frequent visits by people from elsewhere in the world, including the United States and the up and coming Japanese AI-oriented groups.[73] Young researchers were very successful, sometimes gaining an international reputation before obtaining their doctorates: P.J. Hayes, for example, became well known after publishing a joint paper with J. McCarthy in 1968,[74] only receiving his PhD in 1972. G.D. Plotkin's work on inductive inference was considered outstanding;[75] R. Kowalski made a name for himself with his vigorous promotion of the predicate calculus as a programming language in its own right, an approach which became an independent strand of research termed 'Logic Programming';[76] while R.M. Burstall, one of the original members of the Experimental Programming Unit, and Michie's second in command, became established as an outstanding computer scientist with an international reputation in his specialist area — the theory of computation, in which he built up his own group in the 1970s. In 1978, he was appointed to a chair with the title 'professor of AI', despite the fact that computation theory was by that time a general computer science research area, rather than a specialist AI one. The robot project attracted considerable publicity with some five television and film crews visiting:[77] indeed, demonstrations became so frequent as to interfere with the everyday research work and had to be restricted.[78]

However, the concentration of talent, the surfeit of publicity, and perhaps more than anything else, the predominance of research over teaching in AI, attracted hostility from other departments weighed down with heavy teaching responsibilities,[79] and strong pressures grew for the area to normalise its activities. In addition the lack of a career structure for researchers on short-term contracts, coupled with the increasing uncertainty over the future of the centre due to the leadership tensions, led people to start moving elsewhere: Hayes went to Essex; Kowalski to Imperial College; several other researchers to the United States; and Longuet-Higgins himself, along with members of his group, moved to Sussex University. In the course of a couple of years, therefore, many of the most highly respected researchers left Edinburgh, and in some quarters Edinburgh was viewed as being in decline.[80]

The problems facing AI were not restricted to Edinburgh alone, nor was the division of the department the outcome of purely local politics: rather, these events were tied up with national attitudes especially on the part of the SRC. The SRC had never been happy with the breakdown of co-operation over the major robotics grant and had become dissatisfied with the progress made in work on the project. This dissatisfaction stemmed to some extent from a basic lack of sympathy with the goals of those involved in the project. Michie's very ambitious plans for a seven-year industrially oriented programme of research in robotics failed to win favour with the SRC and was never formally submitted.[81] More modest proposals were put forward to maintain the level of effort on robotics, but despite a year's very intensive work on the project, in which programmable assembly using visual recognition of parts was attained[82] (at that time one of the foremost achievements in robotics in the world, comparable with leading work in America and Japan), these proposals were turned down. At that stage, the SRC had become very impatient with Michie, as his entrepreneurial talents did not fit in with their expectations, and the previously mentioned survey of AI by Sir James Lighthill, FRS, Lucasian Professor of Applied Mathematics at Cambridge, and an eminent hydrodynamicist, undoubtedly influenced their decision.

Lighthill's report created a major controversy and was published in April 1973, along with other assenting and dissenting views. Lighthill was highly critical of AI in general and suggested that there were three basic categories of research in the area: work aimed at advanced automation on the one hand, and at computer-based central nervous system research on the other, with in addition a bridge category with the basic component of building robots, which he saw as the essential underpinning for AI to have any claims to unity and coherence. Progress in this category, Lighthill suggested, was virtually non-existent and the building of robots a mistaken enterprise, possibly motivated by a desire on the part of those concerned to 'minister to the public's general *penchant* for robots by building the best they can', and possibly also by 'pseudomaternal' drives to compensate for male researchers' inability to give birth.[83] (It is not hard to detect a reference to Michie's polemical enthusiasm in these comments.) Furthermore, what success there had been, he suggested, was evident only in particular applications and derived from knowledge contributed from the substantive fields modelled, rather than from any AI component. In time, he saw the bridge category as withering away, while work directed towards the two extremes would become integrated with other research in their general areas.

129

Not surprisingly, this caused a major stir in the AI community across the world,[84] and the resulting controversy received much public airing in the press and even on television.[85] Without a doubt, despite Lighthill's protestation that his report

> would simply describe how AI appears to a lay person after two months spent looking through the literature of the subject and discussing it orally and by letter with a variety of workers in the field and in closely related areas of research,[86]

it delivered a blow to the prestige of research in the area from which it has never fully recovered. While Lighthill's comments on robots were directed at the specifically AI category, it appears that they also had some effect on inhibiting robot research and use in Britain in general,[87] whereas in other countries robotics has been a steadily expanding area throughout the 1970s.

In practical terms, the report did affect financial support for research in AI in Britain, particularly in the case of Michie's proposals for robotics research, and it also had some influence on funding in the United States where AI robotic projects were cut back, and the Advanced Research Projects Agency (ARPA), the main sponsor of American work in the area, started insisting on mission-oriented direct research, rather than basic undirected research.[88] These cutbacks took place in the context of the general reduction in public spending in the early to mid-1970s, which affected scientific research in all areas, especially those not seen to be of 'social relevance'. However, the effects were to some extent mitigated in the case of AI: partly by the variety of funding sources supporting the area; partly by the SRC's identification of machine intelligence as an important area of long-range research in its 1972 *Computing Science Review*, which underlined the fact that there was, in any case, no one predominating view on the value of AI; and partly by the expanding nature of computing science in general. Consequently, particular projects were able to get support, especially if their relevance was emphasised and explicit reference to robotics avoided.[89]

Furthermore, the debate over the Lighthill report also led to cognitive science being recognised by the SRC, and a panel was set up to review applications in this area. Thus, there was to some extent a shift in resources to this area, rather than a straightforward cut-back of AI as a whole.

That the reorganisation of AI at Edinburgh, with effective removal of Michie from a central position, was not a purely local affair, but was rather bound up with changes in attitude in the wider scientific establishment, given expression by Lighthill, is borne out by the similar pattern of events occurring at Aberdeen.[90] There, the computer unit was dissolved in 1972 after the SRC refused to renew its grant. Elcock, the organisational prime mover behind the AI interests there (and a colleague of Michie's) had some differences of opinion with the university authorities and the newly established computer science department, in which he was not offered a position to his satisfaction, and left for Canada, where he became involved in building up another AI research group. J.M. Foster, the other senior figure in the computer unit, went to Essex University, where R.A. Brooker, at that time chairman of the Computer Science Department, was in the process of building up AI research interests in an attempt to dispel the non-publishing lethargy prevalent there.[91] However, the situation at Aberdeen never became quite so fraught as at Edinburgh.

In Edinburgh, Michie attempted to fight back against the reorganisation. He marshalled support from the Dalle Molle Foundation to keep research going under his direction,[92] and he had previously published an article in the *New Scientist*, in response to Lighthill's report, asking why it was that AI, which required peanuts in financial terms, should suffer cuts, while nuclear physics, absorbing huge amounts of money and providing little proportionate return, should not be cut:[93] but to no avail. Essentially, by the time of the reorganisation he had lost the support of his colleagues, and Swann, who as Principal had supported and protected Michie, had left in 1973 to take up the chairmanship of the BBC.[94]

In the Machine Intelligence Research Unit after 1974, Michie's energies were channelled into promoting, directing and carrying out research on chess playing programs and organising further machine intelligence conferences, one in the United States with Elcock,[95] one in Russia,[96] and a third, again in the United States. He spent a considerable amount of time on visiting professorships abroad and took up scientific journalism, where in his regular column, 'Michie's Privateview' in *Computer Weekly*, he often commented on the importance of AI research. Latterly, in the late 1970s and early 1980s, he started vigorously promoting 'expert systems', (AI frameworks for representing and mobilising highly detailed specialist knowledge, which, given their essentially simple structure, have achieved remarkable levels of competence comparable to those of human experts), by organising conferences and schools to disseminate the approach to industry,[97] as well as directing and sponsoring relevant research in his unit.

Once the new regime of the Department of AI had settled down, activity did become normalised, with the conventional university emphasis on teaching coming to the fore. This owed much to the mid- to late-1970s financial stringencies which ensured that the universities looked to efficiency in their sphere of production — namely, the training of students. Undergraduate courses were experimented with at Edinburgh, and an AI textbook produced.[98] The application of AI in education itself — intelligent computer-aided instruction, clearly an area of direct social relevance — became a major concern at Edinburgh. This was an area originated by, among others, Seymour Papert, a colleague of Minsky at MIT, and was taken up by Michie and refined and developed at Edinburgh in a variety of approaches by J.A.M. Howe and his colleagues. In general, research activity remained at a high level, producing some 250 publications during the period 1975–80, and with a strong postgraduate school of about 30 being built up after an initial weakening due to the Lighthill report.

In some quarters, however, there was the impression that Edinburgh had declined in importance as an AI centre. This impression was partly due to the departure of highly respected researchers from Edinburgh, as already mentioned, but there were also other contributing factors. Firstly, Edinburgh no longer attracted the publicity over the robotics work as it had formerly done, though research with the robot equipment continued at a modest level; indeed, publicity was shunned as a matter of departmental policy, because of what was felt to have been over exposure by Lighthill. Secondly, an influential PhD thesis by T. Winograd at MIT in 1972[99] had brought the natural language research area to the centre of the AI stage, displacing the theorem proving approach, a shift which owed something to a 'witchhunt' against theorem provers led by Minsky.[100] Edinburgh, however, with a strong tradition in this latter area associated with the research group of Bernard Meltzer, at that time professor of computational logic and head of the department, had no strong competence in natural language and did not appoint a specialist on a permanent basis. Theorem proving had developed a very strong internal theoretical dynamic, deriving from mathematical logic and based on refining the 'resolution' method of machine-oriented inference devised by J.A. Robinson in 1965, which had brought it into the centre of attention in AI the first place.[101] This internal dynamic, coupled with the theorem provers' assured confidence in their formal mathematics-based status, had led to their being always rather autonomous; and under easier funding conditions, they would probably have become a completely independent specialty of computer science, much as pattern recognition, itself based on a strong internal dynamic, had done in the early 1960s.[102] However, the theorem provers made somewhat of a comeback in the late 1970s with the logic programming approach embodied in the language PROLOG, thus re-establishing themselves to some extent as a source of techniques of utility to AI in general.[103]

A third contributing factor to the perceived decline of Edinburgh as an AI centre was that computation theory, another of the major research themes in the department there, under the leadership of R.M. Burstall, professor of AI, had become more central to computer science in general during the 1970s and less of a specifically AI approach.[104] This situation was rationalised in the late 1970s with the transfer of the computation theory group from the Department of AI into the Department of Computing Science. A fourth and final contributory factor to the perceived decline of Edinburgh was that other major concentrations of AI interest had developed in Britain: at Sussex, Essex, and later in the Open University. At the same time, numerous one-person AI projects were pursued elsewhere, often based on 'colonisation' or 'infection' from the established centres.

Professor R.A. Brooker, always an enthusiast for AI (he was one of the panel members in favour of AI in the 1972 SRC *Computing Review*), had wanted to invigorate the research atmosphere in the computer science department at Essex, and to this end had recruited J.M. Foster from the Aberdeen AI research group to a chair in the department.[105] However, this did not work out and Foster left after about a year, having done little on the AI research side, but having developed the elements of a course in the area. There was considerable interest in the AI approach on the part of young researchers in the department such as J.M. Brady and R. Bornat, and when P.J. Hayes arrived in late 1972 from Edinburgh, bringing with him his extensive familiarity with the AI literature and research front, he catalysed the development of research projects in the field at Essex. One such project, supported by the SRC, was the development of a system to read handwritten FORTRAN coding sheets using high-level knowledge to guide the interpretation — a project of clear practical utility (although it never paid off) that nevertheless incorporated the AI approach.[106]

Once established as a centre for AI, other people were attracted there. J.E. Doran, who had been an early member of the Experimental Programming Unit and the Department of Machine Intelligence and Perception in Edinburgh, joined in 1973, bringing with him his research interests in using AI techniques for the reconstruction of cultural evolution in pre-historic settlements from data arising out of the archaeological excavation of graves — yet another illustration of the divergent nature of the AI activity.[107] Bruce Anderson, also from the Edinburgh AI centre, joined for a while before going to the Department of Electrical Engineering, while Yorick Wilks, who made a name for himself with his AI work on natural language and his numerous publications, was appointed to a chair in the Department of Linguistics — marking a new stage in the penetration of other subject areas by AI.[108]

This blossoming of AI interests at Essex was aided by the energetic and aggressive activity of Brady. He rapidly established himself as a competent researcher in the field and became involved at the organisational level, being elected to the chairmanship of the AISB society in the late 1970s, and organising a summer school which attracted leading researchers from the United States. He soon became known in the international AI community and made frequent visits to America, eventually leaving Essex to go to MIT in 1979. The MSc course and research environment at Essex proved to be an effective medium for the training of AI practitioners, some of whom subsequently went on to start AI groups elsewhere — notably H.J. Siekmann, who built up a group in Germany; B. Wielinga in the Netherlands; and C. Bearden, who founded an organisation similar to AISB in New South Wales in Australia. The level of interaction with other centres was high, particularly those in Britain. There were links with Edinburgh through Hayes's connections there, and increasingly with Sussex through exchanges of students. In particular S. Hardy, after doing the MSc course at Essex, went on to do doctoral work in AI and then moved to Sussex, where he was largely responsible for the design of an AI computing environment for the cognitive studies programme.[109]

The emergence of Sussex as a major AI centre in the mid-1970s had a basis that went back to the mid-1960s. N.S. Sutherland, an experimental psychologist who had been interested in the cybernetic and information theory developments of the 1950s, was appointed to a chair at the newly-established University of Sussex. He was favourably disposed towards the AI approach, about which he had heard while he was at MIT in the early 1960s, and was much impressed by the ideas of M.B. Clowes, whom he had met at Oxford and who, as already mentioned, had founded the AISB society. Moreover, AI had been taught by Sutherland as an ingredient of the experimental psychology course from the start at Sussex, and he had arranged for two people from Edinburgh, Burstall and Doran, to visit Essex on a regular basis to give lectures on technical aspects in the area. As part of the conditions for Sutherland coming to Sussex, he had insisted on the founding of a brain research institute, and had negotiated with A.M. Uttley, Gregory, and Longuet-Higgins, for them to join him.[110] The latter two decided to go to Edinburgh instead, but Uttley came to Sussex and started work on the simulation of networks of an artificial neuron (the informon),[111] and on the application of these to perception. This was clearly a cybernetic rather than an AI project, but the computer provided for the project by the SRC made possible a subsequent characteristically AI attack on machine vision, which was carried out by Clowes when he returned from Australia in 1969 — work which rapidly established him as a leading figure in machine vision research.[112]

At that time in Sussex there was a broad base of interest in AI, favoured by the explicit focus on interdisciplinarity of the distinctive Sussex 'school' organisation which contrasted with conventional departmental divisions and their associated impervious boundaries,[113] while Asa Briggs, then vice chancellor of the university, was supportive of new ventures.[114]

In 1970 Sutherland put forward a radical proposal for a new School of Cognitive Studies with an intellectual focus on

135

knowledge and understanding to include teaching and research in a range of subjects: computing science and AI; experimental psychology; linguistics; logic and philosophy; and mathematics.[115] However, in the restrictions on growth of the universities in the early 1970s, this proposal was turned down. However, a more modest development, which came to be called the 'Cognitive Studies Programme' was eventually launched within the existing School of Social Sciences, and in 1973 Clowes moved from experimental psychology to take up a chair in AI instituted for the programme. Several other members of the university were associated with the programme, among them M.A. Boden who had become familiar with the computational approach during her doctoral research in the United States on purposive behaviour in psychology,[116] and whose book *Artificial intelligence and natural man*[117] is one of the most accessible introductions to work in the area; and Aaron Sloman, a philosopher with a mathematics and physics background who was influenced by Clowes and Boden to consider work in AI and who spent a year in the Edinburgh centre before returning to start research in vision at Sussex.[118] His book, *The computer revolution in philosophy*[119] argues enthusiastically for the great potential of the AI approach in matters philosophical. Meanwhile, in 1974 Longuet-Higgins and some members of his research group from Edinburgh had joined the Centre for Research in Perception and Cognition, a research unit associated with the Laboratory of Experimental Psychology. Following Longuet-Higgins's move, other members of experimental psychology, Professor P.N. Johnson-Laird and C. Darwin, also developed research interests in AI,[120] which was taught as a compulsory part of the course presented there.

These two AI-oriented groupings ensured that Sussex emerged as a major centre for AI, especially in the areas of language studies and vision, in the 1970s. Moreover, the emphasis on the 'soft' applications of AI — the study of cognition, rather than on the 'hard' areas like the computer science and engineering applications — marked the rise to importance of cognitive science — the computational approach to linguistics, psychology, and philosophy, based primarily on the methods of AI. This rise was encouraged by the formation of a Cognitive Science panel in the SRC. Similar cognitive science concentrations emerged elsewhere, usually as a supra-departmental federation for research, as in the Edinburgh

School of Epistemics (founded by Longuet-Higgins and others in 1969, but only having really taken off in the late 1970s) and the Cognitive Science Institute at Essex, also implemented at the end of the 1970s. At the Open University similar developments had occurred. The Cognitive Psychology course included a substantial AI component, while the presence of other researchers, interested in the application of AI to education, meant that the Open University itself comprised an emerging centre for AI in the late 1970s.

As well as being an essential element in the emergence of cognitive science, AI has become a recognised specialty within computer science.[121] While these developments in the 1970s bear some resemblance to Lighthill's predicted fission of AI research between the categories of computer-based central nervous system research on the one hand, and advanced automation on the other, it is difficult to align his prediction with the continued coherence of AI — that is, the continued identifiable existence of the AI paradigmatic structure. Rather than the established areas of linguistics, psychology, and philosophy absorbing AI, it would seem that cognitive science is an emerging synthesis based on the unifying computational modelling approach of AI. Indeed, one could equally well argue, against Lighthill, the AI practitioner's extreme view that what is happening is merely the process of colonisation of other areas by the AI approach: 'I see the future of AI as a very long haul towards computational theories of physics, chemistry, linguistics, sociology, visual perception, locomotion and every other aspect of what it means to be human'.[122] Clearly, therefore, views still differ over the assessment of the place and future of AI.

3.7 THE ESTABLISHMENT IN THE UNITED KINGDOM

One of the main features of the development of AI in Britain was the initial and continuing strong American influence. Nearly every one of the leaders of AI research in Britain had visited the United States and been impressed by developments there, before moving into the area, or promoting it themselves in Britain: Michie, Meltzer, Sutherland, and Clowes had all visited AI projects in the United States in the early 1960s. These links with the American AI community were maintained and strengthened in the ensuing years, and continue today with a high international exchange of personnel between the various centres.

These links also underlie the substantive similarity of research pursued in Britain and America, at the general level of the paradigmatic structure, as reflected for instance in the research profile at Edinburgh which matched the patterns evident in the United States. This has remained true since the mid-1960s when AI took off in Britain, and it has largely been the case that the initiative in the development of AI in terms of the broad content of research has remained in the hands of the establishment in the United States. Consequently, the emergence of the establishment in Britain has been bound up with organisational aspects of development to a greater extent than in the United States, where organisational and general substantive innovation were both important. Nevertheless, just as in America, the emergence of the establishment in Britain has been inextricably bound up with the development of the field. The organisational aspects of development have not only predominated in Britain but have also had an appreciable impact on the shaping of the wider international AI community: for instance, the Machine Intelligence Workshops played an important role leading to the establishment of a journal and an international conference structure for the area; the first dedicated AI department was established at Edinburgh; and the cognitive science concentration has received its firmest institutional expression in Britain, with the Cognitive Studies Programme at Sussex, and the School of Epistemics at Edinburgh.

Moreover, in the development in Britain, a clear division between what have been called organisational and intellectual leadership roles[123] is identifiable, especially with respect to the first generation establishment. At Sussex, for instance, N.S. Sutherland was energetic in supporting the AI approach there, getting grants for researchers to work in the area without himself being actively involved. Similarly, at Essex, R.A. Brooker deliberately encouraged the development of an AI research group within the Department of Computer Science, but did not himself actively contribute to research to any great extent. It was also the case at Edinburgh with Michie, whose contributions in substantive terms were overshadowed by his role as an organisational leader: indeed, he ranks as a scientific entrepreneur of the first order. He alone, almost single-handedly, was responsible for getting AI launched in the United Kingdom, and his influence appears to have lurked behind nearly every event of major importance concerned with AI in Britain in the 1960s.[124] While his enthusiastic promotion of robotics was eventually to backfire with Lighthill's condemnation, his activity was nevertheless instrumental in putting Edinburgh on the map with respect to AI, and for providing an environment in which young researchers were able to establish international reputations. Despite his fall from favour after the Lighthill report, Michie remained active — organising a further three Machine Intelligence Workshops; promoting and directing research on chess-playing programs; and, latterly, working to bring the 'expert systems' applications area of AI to the notice of a wider audience, including industry. Also at Edinburgh, Meltzer established himself as an 'elder statesman' of AI, channelling his energies to maintaining high critical standards in the specialty Journal, and to providing an environment in which major developments in theorem proving were made; while his own substantive contributions have been in the form of synthesising reviews, directing attention to certain problems and suggesting fruitful possibilities. Even Longuet-Higgins, with his dislike for organisational involvement, preferring purely scientific activity, has played a role of organisational importance, in arguing for the cognitive science applications of AI, and in helping to found the School of Epistemics at Edinburgh.

An important part played by the organisational role has been the provision of facilities and opportunities for young researchers and students to rise into the establishment, in many cases developing an international reputation for themselves with their PhD work. The high number of important contributions to the field at PhD level seems to have been characteristic of AI, and has combined with the very fluid and informal nature of the organisation of work in the area, to maintain a shallow internal hierarchy with no elaborate vertical division of labour. Terry Shinn's description of the organisational structure in laboratories concerned with computer research in vector analysis could apply equally well to the case of AI.[125] While this organisational fluidity could stem to some extent from the holistic and diffuse nature of the research goal of modelling human intelligence, it seems likely that the situation also arises out of the youth of the field. The preponderance of young researchers at about the same stage in their professional careers, and hence of more or less equal status, together with the general shortage of personnel with the appropriate computer background, has rendered it difficult to support a steeply hierarchical structure — even where such a structure *would* be possible, as, for example, in some well routinised commercial computing where there has already been a drive towards extreme stratification and division of labour along scientific management lines.[126] However, there has been a horizontal division of labour, with specialisation in different research areas and, of course, organisational roles have tended to be filled by professionally senior practitioners. It is likely that the taking up of such roles by the first-generation establishment was the outcome of two factors: on the one hand, the clear opportunities offered by these roles and their availability in the early days of the emergence of AI in Britain; and on the other hand, the difficulty of getting to the research front without extensive specialist training, particularly in the activity of programming, at that time very much an esoteric art. Moreover such training obviously required some sort of organisational structure for efficient transmission. Consequently, the organisational roles were more important and probably demanded more effort during the initial emergence of AI in Britain, before they became institutionalised and suitable recruits were readily available from among the ranks of the trained AI specialists. With the development of the institutional structure during the 1960s and 1970s, there has emerged a similar tight pattern of intergenerational and intercentre linkages in Britain as that evident in the United States: indeed, the American and British structures were connected, as is illustrated by Figure 3.2.

This tight intergenerational pattern has been reinforced by two related characteristics of the AI paradigmatic structure: the *constructive* and the *craft* nature of work in the area. The writing of a computer program to carry out some task clearly involves making or constructing something, rather than investigating something that is naturally given (although such investigations can be and are involved). Not only are there many ways of constructing a program on AI principles, but there are also many other non-AI ways — for instance, the construction of a stochastic model. Furthermore, it is possible to construct models not based on the use of the digital computer — for instance, in the building of electronic analogs of neuron

Figure 3.2: Movement of research personnel between AI centres in the United Kingdom, 1964–80. Each line shows the trajectory of an individual; ●—— represents the start of a career in AI. The density of horizontal lines gives some indication of the frequency of movement between centres. The figure represents trajectories for 50 out of a total of about 150 personnel involved in AI research in Britain during the period 1964–80; a further 43 did not move, and 15 left the field altogether. Information was not available in the remaining 40–50 cases, and is less complete after 1978, while short-term visits of six months or less (which were quite frequent) have not been included. Thus, the above is probably an understatement of movement between centres. The following features are apparent: strong links with the United States; movement away from Edinburgh in the mid-1970s; increased movement in the early to mid-1970s involving the newly emerged centres at Sussex and Essex; increasing links with centres elsewhere in the 1970s; and finally, the importance of Edinburgh, and latterly Essex, as sources of AI personnel.

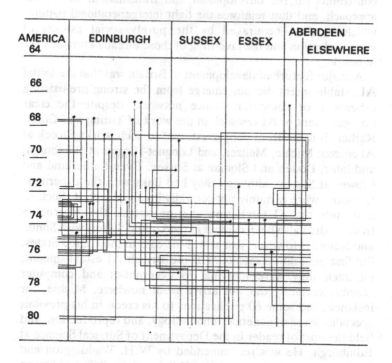

141

networks. Hence, it is clear that the AI paradigmatic structure delineates only one broad way of making models among many possible alternatives; this arises from the constructive nature of the activity in the area. Central to the paradigmatic structure is the activity of programming, based on the use of list processing languages, associated with which are many characteristic techniques and 'tricks of the trade'. Effective programming also involves a high degree of skill, and all of these features together lend the AI approach a distinctly 'craft' nature, which requires for its transmission a lengthy apprenticeship and some degree of contact and interaction with experienced practitioners. This craft and constructive nature of work in the area ensures that it is extremely unlikely that the specifically AI approach, even in broad terms, would be developed spontaneously and independently outside the community of people already using it. Consequently these aspects of the paradigmatic structure place constraints on the development and transmission of the AI approach, and thus reinforce the tight intergenerational pattern, which is also encouraged by the purely social aspects of communication and the favouring of those already known in the network.

A major feature of development in Britain was that the initial AI establishment did not emerge from the strong pre-existing cybernetic or computer science network, despite the clear prefiguration of AI research in the work of Turing and Craik. Rather, it came from people external to such work: Elcock at Aberdeen; Michie, Meltzer, and Longuet-Higgins at Edinburgh; and later, Boden and Sloman at Sussex. Even Sutherland and Clowes at Sussex, although they had links with the cybernetic tradition, were certainly not centrally involved with it. Such an entry into the AI establishment from outside the cybernetics tradition had also been evident in the United States, with Simon and Newell. However, just as was the case in the United States, the first generation members of the British AI establishment, although they were marginal to cybernetics and computer science, certainly did not arrive from nowhere. Michie, for instance, had some 60 publications to his credit in his previous specialist areas of genetics, immunology, and reproduction, and held the post of reader in the Department of Surgical Science at Edinburgh. He was recommended by W.H. Waddington and M. Swann on his appointment to Edinburgh in 1958, and Swann continued to support him during his AI activities. Meltzer, as a reader in the Department of Electrical Engineering, also had a solid reputation for his work on electron beam dynamics (used by NASA for the design of ion propulsion for space vehicles) and solid-state electronics; and Longuet-Higgins, with his international standing in theoretical chemistry was clearly already a member of the wider British scientific establishment.

The migration of outsiders with some standing, and hence the freedom to move fields,[127] therefore seems to have had as its major consequence the construction of an organisational structure within which the pursuit of AI research subsequently developed. In some cases (especially that of Sutherland) the organisational leadership role merged with another, probably necessary, role in the emergence of a new interdisciplinary area — namely, a sponsorship role. The institutional developments at Edinburgh would not have been possible without some strong support from sponsors placed in fairly influential positions with the university government, especially in view of the tendency for the publicity attracting, research intensive, AI activities to arouse suspicion and resentment: both Sir Edward Appleton and Sir Michael Swann, who succeeded Appleton as Vice Chancellor of Edinburgh University, were active in encouraging and supporting those developments; and at Sussex, there was also fairly widespread support for AI-oriented developments among many of those in positions of influence.

As well as this positive sponsorship within the wider establishment, which was accompanied by a positive evaluation of the status of AI, there was a negative sponsorship as well, as demonstrated by Sir James Lighthill's report. Lighthill's opposition could clearly be seen to support the *status quo*: he affirmed the value of the currently existing areas included in advanced automation (e.g. control engineering) and in central nervous system research (e.g. neurophysiology), and moreover he attributed any success in AI to contributions arising from these areas. The validity of the emerging interdisciplinary area of AI was thus challenged and denied, explicitly in terms of the already-established disciplines surrounding AI. In particular, Lighthill's focus on the established central nervous system areas of research — neurophysiology and neurochemistry — and his use of the term 'central nervous system' happened to align with the dominance in Europe of the neurosciences which study the 'hardware' of the brain, over the cognitive sciences — linguistics and psychology — which might be said to study the 'software' of the brain, a dominance which is not as clear-cut in the United States.[128] In this dominance relation we see a prestige hierarchy, with those sciences closest to the physical sciences accorded most prestige. In this context, the study of the brain at the reductive level of biochemical or neurophysiological mechanisms is considered more prestigious than the AI approach at the information-processing levels.

Moreover, it is difficult to account for the impact of the Lighthill report (by his own admission a two-months layman's view of the area)[129] except in terms of the authority carried by Lighthill's eminence. It is interesting to note that reference is still offered to observations that Lighthill made on the difficulties for search arising out of the combinatorial explosion,[130] as if he were their originator, whereas, in fact, the combinatorial problems had been regarded as the *raison d'être* for AI — the huge size of the space of possible moves in chess, for example, estimated by Shannon in 1950 as in excess of 10^{120}, dictated the need for heuristic strategies to restrict the search space to manageable proportions.[131]

What is also of interest about the Lighthill report, however, apart from its importance as an authoritative pronouncement on the status of AI research, is that it is one among a multitude of attacks on the field.[132] Such attacks on AI have been commonplace, and while they purport to deal with the particularities of the subject matter of research in the area, it is quite clear on closer inspection, that they are more concerned with the general goal of constructing an intelligent machine. It would be too lengthy to argue this fully here, but it is perhaps pointed by the continued relevance of Turing's comments on the arguments for and against the possibilities of constructing intelligent machines,[133] despite the fact that the distinctive AI approach had not emerged when Turing was still alive. It is also pointed by the fact that these attacks on AI are not the prerogative of any particular group: criticism has come from all shades of political opinion, and from all areas of research, scholarly as well as technical.

Ironically, this variety in attacks can only be matched by the diversity in the sources of support for AI, or the range of (often conflicting) views within the field itself. Something of this has already become evident in the differences between Longuet-Higgins and Michie, one favouring the cognitive science definition of AI, and the other the machine intelligence approach. There are other divisions: those supporting a theoretical formal approach, such as McCarthy, for instance, and those supporting the exploitation of practical applications, such as Feigenbaum; those who see no problems with accepting military funding (McCarthy and Feigenbaum) and those implacably opposed (Meltzer, while Michie was opposed to classified work).

These divisions, which are legion in the area, are coupled with a rather amazing state of substantive partisanship, or scientific ethnocentricity, in which proponents of the various different research areas each tend to see their own approach as the real AI approach — to the theorem provers, theorem proving is central; for the natural language proponents, language is the basis for reason, and so on.[134]

Many of these differences can be related to the background competences of the practitioners, and can be interpreted as competition between groups on the research area level for resources and authority within AI as a specialty, constituting perhaps the primary locus for competition over cognitive commitments, quite distinct from the individualistic level of competition for recognition, long identified in the sociology of science as a basic motor of scientific development. An important point to note is that such differences are not precluded by the paradigmatic structure of AI, outlined earlier as providing guidelines for the common computational approach and its programming basis, but which does not dictate a dogmatic monolithic attitude, nor inculcate a unifying solidarity. Moreover, this variety of views in and around AI can be related to its position as an interdisciplinary area, with particular research areas associated with particular neighbouring disciplines — for example, the natural language research area is associated with linguistics, while theorem proving has links with metamathematics; and as an interdisciplinary area, the status of AI research is still very much in process of negotiation. The cognitive science developments appear to have led to an acceptance, on the part of those involved of the validity of AI: indeed, the impression in that context is that AI is the 'hard' formal core, and therefore of high status. However, in a computer science or general scientific context, AI is still seen very much as a 'freaky', rather dubious fringe activity, and consequently of rather inferior status.[135] Moreover, there are two broad categories of attacks which can be related to these contexts. On the one hand there are attacks, often by philosophers and others, on AI for being reductionistic and impossible[136] — in a sense it is 'harder' than is appropriate for the study of intelligent activity. On the other hand there are criticisms, often by computer scientists, of AI for being morally wrong,[137] bad science,[138] or undisciplined and sloppy.[139]

Finally, perhaps the variety and depth of feeling of the many attacks on AI derive not so much from what is in fact done in AI research, but rather from the fact that the very broad aim of research in the area — namely, the construction of intelligent machines — bears uncomfortably on our conception of ourselves. In Elias's terms, AI research is seen to be involved with a very sensitive area of the means of orientation: the area which is concerned with the nature of mind. Furthermore, during several hundred years of development and struggle with other competing establishments, a scientific establishment has yet to succeed in gaining a monopoly over the means of orientation in this area. In making its challenge in this area, therefore, AI is inviting violent attacks, and its practitioners should hardly be surprised when they suffer them.

3.8 CONCLUSION: THE PROCESS OF DEVELOPMENT AND ESTABLISHMENT IN ARTIFICIAL INTELLIGENCE

The process of development and establishment in AI in broad terms therefore appears to have been as follows. Around the period of the Second World War, catalysed by the war-time weakening of traditional disciplinary boundaries, and brought into being by exigencies deriving from the unprecedented problems of organisation and communication posed by the increasingly complex social structures and conditions, there emerged the software sciences. These had their focus on pattern rather than substance, and included operations research, computer science, and cybernetics. In particular, the cybernetics area, with its focus on the processes common to animals and machines, promised a realisation of the age-old desire to make an artificial human, a machine that could think. Within the general area of cybernetics, various approaches were made to the construction of intelligent machines, some based on electronic analog of neuron networks, others on the simulation of processes by means of the newly developed digital computer.

146

In this context, the paradigmatic structure of AI was articulated in the United States during the late 1950s.

The American establishment in AI consisted essentially of those who had contributed to this articulation, and who had provided the institutional structure within which subsequent research, based on the distinctive AI paradigmatic structure, could be undertaken. The source of authority and reputation of the establishment was derived from the effective demand which developed for this distinctive approach on the part of those who wanted to follow the approach themselves and those who thought the approach was worthwhile and promising. In addition, the emerging establishment secured the backing of the funding agencies, aided by their good connections with the wider establishment, in competition with other approaches within computer science. The preference of the funding agencies for concentrating resources in a few centres, coupled with the expense of the instrumental base required (the digital computer), ensured that an effective monopoly over material resources as well as the cognitive ones was maintained.

As the paradigmatic structure was already elaborated before it was exported to the United Kingdom in the early to mid-1960s, it left less opportunity for substantive contributions on this general level while there was ample scope for extending the institutional facilities for carrying on AI research in Britain: consequently, the first-generation British establishment in AI consisted of those who were able to set up organisational forms to exploit the already articulated paradigmatic structure, thus giving them a monopoly over the cognitive resources in the area. Because of the constructive nature of AI, which was manifested in it being only one among several competing cybernetic approaches, and since people already committed to a particular approach tend to stay with that approach, members of the first-generation establishment included people from outside the cybernetic and even computer science traditions. This was the case in the United States, but was more marked in Britain. Moreover, the members of the first-generation establishment were, in fact, drawn from those who already had some standing or prestigious backing in another field; and being thus well connected, they were able to secure the backing of the Science Research Council in competition with other approaches within computer science, ensuring their effective monopoly over material as well as cognitive resources in the area.

This monopoly over resources was reinforced by the constructive and craft nature of the AI paradigmatic structure, and the patterns of development in both the United States and the United Kingdom followed the lines of personnel mobility and contact, thus leading to a tight intergenerational and intercentre structure of linkages. In particular, students and 'descendents' of the members of the first-generation American establishment have tended to dominate the field. The paradigmatic structure of AI, however, while providing general guidelines for the methodological approach employed in AI research, does not dictate the direction of research. Due to the very wide-ranging nature of the focus on intelligent activity, research in AI has become involved with the subject matter of many other disciplines, and has therefore developed as an interdisciplinary area. This interdisciplinary character of AI has induced many mutually competing divisions *within* the area, as well as leading to many external views on the status of the field, which has consequently been very much a matter for negotiation: from the point of view of the 'soft' sciences, such as linguistics and psychology, AI has appeared 'hard' and therefore of superior status; from the point of view of the 'hard' sciences, such as computer science and physics, AI has appeared somewhat 'freaky' and therefore of inferior status. Furthermore, as a newly emerging specialty, AI has been in competition with already established disciplines: the Lighthill report, critical of AI, can be interpreted in this context as an affirmation of the *status quo*. Finally, because of the general aim of constructing intelligent mechanisms, AI has been seen as challenging the monopoly on the means of orientation with respect to the nature of mind. As this bears directly on peoples' conceptions of themselves, deep feelings have been aroused, as is evident in the many and varied attacks on AI.

It is therefore evident that there has been competition on a variety of levels over AI, with differing consequences for the development of the field, and, moreover, engaging distinct groups or establishments. At the most circumscribed level, *within* the field, there has been competition over the choice of techniques to be used in a particular research area, and competition between research areas themselves. This is of immediate consequence for research at the practical level, and involves groups negotiating for ascendency within the AI establishment. At a less circumscribed level, the establishment of AI as a whole has been involved in competition with neighbouring scientific and scholarly establishments. This concerns the general scientific validity of the field rather than being of immediate practical import, and has consequences in broad terms for the status of the field, the availability of funding, and the continued demand for the AI approach on the part of potential entrants. Finally, at the most general level, and most clearly underlining Elias's remarks about the means of orientation, the validity of the AI approach is discussed in terms involving political, religious, moral, philosophical, and cultural issues. The question being asked here is not whether AI is valid as a scientific approach, but rather whether any such approach to the mind is viable. This engages a far wider group, and the prevalence of the debate at this level is perhaps pointed by the fact that, at most, a couple of dozen full-time occupational opportunities are available in AI — very few compared with many other areas of endeavour — while nearly everyone, it seems, has something to say on the issue of whether machines can think.

3.9 POSTSCRIPT:
THE COMMERCIALISATION OF ARTIFICIAL INTELLIGENCE

Since the foregoing was written in 1980, there have been dramatic developments in the general reception afforded to AI, with large increases in public funding for research in the area, and even larger increases in commercial interest and financing. Indeed, the early 1980s marked the transition from a phase of 'establishment' to one of 'commercialisation', in which attempts were made to convert or 'transfer' (the fashionable term) the results of AI research from the realm of the merely fascinating into commercially useful products which could be sold, it was hoped, for a profit.

This recent phase of development was triggered by the news that the Japanese Government planned a large-scale programme to put Japanese industry into the forefront of world computer developments, an area where, unlike many others such as manufacturing, the Western nations still had a significant lead.[140] The *Summer 1980 JIPDEC Report*, (JIPDEC: Japan Information Processing Development Centre) put it bluntly:

. . . the Ministry of International Trade and Industry initiated its Fifth-Generation Computer Development Project, a project which, with a target year of 1990, has the aim of developing truly world leading computer system technology and promoting the development of Japan's computer industry through research and development into fifth-generation machines.[141]

This report was published (in English) in September 1980, and outlined the Japanese plans, emphasising the need for supporting and exploiting the results of AI research, especially natural language and knowledge engineering work. These plans were heralded with great excitement, even trepidation: for instance, in the United Kingdom, Alex d'Agapeyeff, an eminent computer software entrepreneur,[142] and at one time a president of the British Computer Society, claimed at a meeting in London in December 1981 that the Japanese were declaring 'economic war' on the West; and that the language they used was that of *Mein Kampf*.[143] D'Agapeyeff went on to play a further part in mobilising concern, by conducting surveys and organising other initiatives.[144]

There were similar reactions elsewhere. In the United States, Edward Feigenbaum, one of the leading 2nd generation American AI practitioners,[145] and Pamela McCorduck[146] collaborated to produce a book about these developments, which called for a concerted American response:

America needs a national plan of action, a kind of space shuttle program for the knowledge systems of the future. In this book we have tried to explain this new knowledge technology, its roots in American and British research, and the Japanese Fifth Generation plan for extending and commercialising it. We have also outlined America's weak, almost nonexistent response to this remarkable Japanese challenge. The stakes are high. In the trade wars, this may be the crucial challenge. Will we rise to it? If not, we may consign our nation to the role of the first great postindustrial agrarian society.[147]

This book became a best seller, selling 10,000 copies within a matter of months in the Japanese translation alone.

The Japanese Ministry of International Trade and Industry

(MITI) announced the commencement of their ten-year, US$850 million effort in October 1981, and other national governments were not long in responding with comparable initiatives. In the United Kingdom, the Alvey programme was started in 1983, following the recommendations put forward in a 1982 report, *A programme for advanced information technology*,[148] by a study committee set up as a more or less direct response to the Japanese plans. Funding of some £352 million was proposed, of which £26 million was to be spent directly on Intelligent Knowledge Based Systems (essentially relevant AI research), with more for AI through the £78 million funds for 'demonstrators' and education. As Patrick Jenkin, the Secretary of State for Industry, said, in a statement to the British House of Commons:

> This is the first time in our history that we shall be embarking on a collaborative research project on anything like this scale. Industry, academic researchers and Government will be coming together to achieve major advances in technology which none could achieve on their own. The involvement of industry will ensure that the results as they emerge are fully exploited here in Britain to the advantage of our economy. Information technology is one of the most important industries of the future and therefore one upon which hundreds of thousands of jobs in the future will depend.[149]

In the United States there were several comparable programmes, including the Microelectronics and Computer Technology Corporation (MCC), and the Defense Advanced Research Projects Agency $1 billion Strategic Computing programme. The former was officially launched in January 1983, after nearly a year of meetings, while the decision to go ahead seriously with the latter was made in spring of 1982,[150] following the news of the Japanese initiative. Many other relevant programmes have also now started around the world, notably the multi-nation European Strategic Programme for Research and Development in Information Technologies (ESPRIT), and the controversial American Strategic Defense Initiative (SDI), popularly known as 'Star Wars'.

Clearly, therefore, the climate of reception for AI in the 1980s changed considerably from the sceptical years of the Lighthill report. Catalysed by the innovative example of the

Japanese intiative, information technology was placed firmly on the agenda of every major government, with AI at the centre of attention and effort.

In the UK, this resulted in a large growth in the numbers of people interested in AI, and in employment opportunities in the area. At the end of the 1970s, there were perhaps a dozen or so full-time practitioners, with approximately another hundred people with significant interest; by 1986 there were probably over one hundred full-time AI posts (though not necessarily occupied by fully trained or experienced AI practitioners!), over one thousand people with significant interests, and many more sufficiently interested to attend the increasingly expensive conferences on the subject.[151] From the handful of UK academic institutions involved in the 1970s, by the mid-1980s nearly every institution of higher education in the UK showed signs of serious interest in AI research and training.[152] By the mid-1980s also, substantial industrial and commercial involvement was developing, with many companies seriously looking at the potential for making use of AI systems in their own operations.[153] An AI supply and service infrastructure was emerging from what had been a very small basis indeed in the 1970s, with perhaps three or four companies supplying AI products of some sort: by 1986, there were some 24 new start-ups (and at least two close-downs), and in addition, many major established electronics and computer firms were developing in-house AI divisions.[154]

The basic patterns of research in AI outlined earlier (in section 2 of this chapter) and in existence by the early 1960s, as discussed in section 4, were still essentially recognisable in the late 1980s. Indeed, it became routine to talk of AI 'tools'[155] while the research area structure continued to develop and differentiate. There were, however, perhaps three major changes affecting the development of the substantive structure of the area, which came to the fore during the 1980s.

The most obvious was in the rise of the so-called 'expert systems' area, with its emphasis on 'knowledge engineering' or knowledge-based information processing. This was seen to be the leading edge as far as practical commercial exploitation was concerned, and attempts were made at building expert systems to deal with a wide range of practical domains.[156] It was also the area identified by the Japanese as being at the core of the 5th generation computer systems.[157] Essentially what happens in the development of an expert system is that AI knowledge-representation methods and techniques are used to capture the knowledge of a human expert in the target domain, to produce a knowledge base. This base is in turn processed by other AI techniques for making inferences, to draw appropriate conclusions as required, and thus emulate the intelligent reasoning of the original human expert. The development process for expert systems requires close interaction between the AI practitioner or 'knowledge engineer', and the domain expert.

Another major change, although not one with immediate or direct implications for the commercialisation of AI, was the rise of Cognitive Science. This referred to the use of the computational model in the human sciences, often in a purely metaphorical sense rather than necessarily implying the implementation of a computer program, and it has been argued that this represents a paradigmatic shift for the human sciences.[158] Given the massive attention afforded AI through the developments outlined above, the exploitation of computational ideas in this way was hardly surprising, and compares with the similar exploitation of conceptual resources arising from previous 'new' technologies such as hydraulics or clockwork mechanisms.[159]

The third major change affecting the structure of AI, and one which in fact underpinned and supported the other two changes, as well as underlying the broader field of information technology in general, was the dramatic decrease in computing costs. This lowered the hardware resource barrier to carrying out AI research, a barrier which had been very much in evidence during the early years of development, as previously discussed. As a result, the scope for a strong AI establishment to dictate and control development in the area was considerably weakened, allowing outsiders to move in and pursue their own variants of AI research.[160] Low computing costs should also encourage the rapid and wide diffusion and use of expert systems and other AI products, once their viability and efficacy has been demonstrated. Indeed, it became possible in the early 1980s to buy an expert system 'shell' (that is, the basic structure of an expert system, without the domain specific elements) for a personal computer, although the general efficacy of such systems appears rather constrained.

However, despite the highly favourable climate and the dramatic growth of effort in AI, debates over the validity of the approach and associated claims remained as lively as ever.[161]

Indeed, it could be argued that the far bigger market of interest in AI provided an even better basis for making a reputation for oneself by producing a critique of the area.[162] AI was still as popular as ever a focus for attack: clearly the struggle for mono-polisation of the means of orientation continued unabated.

It was also not clear, at least by the mid-1980s, that practical commercial success had been consolidated or even proven, whatever the promise of research or prototypes. Some systems, in actuality constructed along more or less conventional lines, have been somewhat misleadingly described as expert systems, while in any case opinions differ widely over what really constitutes an expert system, and over just how many are in existence.[163] Even firms in the emerging supply and service infrastructure appear to make most of their sales to other suppliers and research and development teams, rather than to a wide base of satisfied users.[164] Furthermore, despite the increases in funding and personnel, it was also not clear that research results had benefited, at least not by the mid-1980s. A survey for one of the Alvey monitoring efforts (set up to monitor progress on the Alvey programme), which looked at publications in AI, found that while the total number of items published increased dramatically, this increase was almost totally accounted for by review articles rather than by more substantial pieces of work.[165]

However, this situation should not be surprising, in the light of the identification of the tight intergenerational craft-constrained nature of development in the area, as discussed earlier in the chapter. At the core of this process of development is a high degree of what might be called 'taciticity', by which I mean that the craft knowledge and practices which constitute the core burden of AI expertise have not yet been made sufficiently explicit, nor sufficiently generalised from the contingencies of their development, to be readily transferable without extensive hands-on experience or appreciable apprenticeship periods: in short, they are still largely tacit. If this taciticity hypothesis is valid, then, given the very small numbers of fully experienced or trained personnel in existence at the end of the 1970s, and given the demands on those people to carry out all sorts of tasks such as research and industrial consultancy as well as training, it is clear that significant effort will be slow to build. Meanwhile, most of the effort by the community as a whole needs to be taken up with learning;[166] and what better way of learning about a new area than writing reviews?

The question nevertheless remains whether AI is in fact yet sufficiently mature for such whole-scale exploitation, as it is in scientific terms a very young field, and, moreover, one which is attempting to tackle extremely difficult and profound problems. The doubt will surely linger that, while the hot-house climate of commercial exploitation and abundant fertiliser of industrial funding will bring on certain exotic fruits, others, perhaps more subtle or sensitive, and possibly more rewarding in the long term, will suffer.

ACKNOWLEDGEMENTS

I should like to acknowledge the help given by the interviewees to whom a draft of this paper was circulated. The comments of J.A.M. Howe, H.C. Longuet-Higgins, D. Michie and N.S. Sutherland were particularly valuable. I should also like to thank R.D. Whitley for his suggestions and the SRC for its support. Finally, thanks are due to the ESRC for support relevant to the writing of Section 3.9, the postscript.

NOTES AND REFERENCES

1. The term specialty is used here in the sense of a community of practitioners differentially located with respect to a common paradigmatic structure which defines a general focus of attention. Compare R. Whitley, 'Components of scientific activities, their characteristics and institutionalisation in specialities and research areas' in K. Knorr, H. Strasser, and H.G. Zilian (eds), *Determinants and controls of scientific development* (Reidel, Dordrecht, 1975), pp. 37–73.

2. See, for example, I. Aleksander, 'Artificial Intelligence', *Electronics and Power*, vol. 22 (1976), pp. 242–4.

3. This paper is based on a detailed MSc/PhD study of the area: J. Fleck, 'The structure and development of artificial intelligence: a case study in the sociology of science', unpub. MSc diss., University of Manchester, 1978. Also PhD diss. in preparation.

4. Compare R. Whitley, 'Components of scientific activities' (fn. 1).

5. Sandewall has discussed these aspects, for example, E. Sandewall, 'Programming in an interactive environment: the "LISP" experience', *ACM Computing Surveys*, vol. 10 (1978), pp. 35–71.

6. Compare J. Law, 'The development of specialties in science: the case of X-ray protein crystallography', *Science Studies*, vol. 3 (1973), pp. 275–303.

155

7. This is clear from a consideration of the contents of the book, a classic in the area, E.A. Feigenbaum and J. Feldman (eds), *Computers and thought*, (McGraw Hill, New York, 1963).

8. Compare D.E. Chubin and T. Connolly, 'Research trails and science policies: the shaping of scientific work by hierarchies and elites', in N. Elias, H. Martins and R. Whitley (eds), *Scientific establishments and hierarchies*, Sociology of the sciences, vol. VI (Reidel, Dordrecht, 1982).

9. D.O. Edge and M.J. Mulkay, *Astronomy transformed: the emergence of radio astronomy in Britain*. (Wiley-Interscience, New York, 1976).

10. Though for a dissenting view, see B.R. Martin, 'Radio astronomy revisited: a reassessment of the role of competition and conflict in the development of radio astronomy', *Sociological Review*, vol. 26 (1978), pp. 27–55.

11. See E. Yoxen, 'Giving life a new meaning: the rise of the molecular biology establishment', in N. Elias, H. Martins and R. Whitley (eds), *Scientific establishments and hierarchies*, Sociology of the sciences, vol. VI (Reidel, Dordrecht, 1982).

12. N. Elias, 'Scientific establishments', theme paper for Conference on Scientific Establishments and Hierarchies, Oxford, July 1980.

13. N. Wiener, *Cybernetics — control and communication in the animal and machine*, (Wiley, New York, 1948).

14. For instance, W. Pitts and W.S. McCulloch, 'How we know universals', *Bull. Maths. Biophysics*, vol. 9 (1947), pp. 127–47; or O.G. Selfridge, Pandemonium: a paradigm for learning' in *Mechanisation of thought processes* (HMSO, London, 1960), pp. 513–26.

15. Machine Translation was an early example of this approach. Based on syntactical analysis and dictionary look-up, it failed at that time in its aim of providing high quality translation — see Y. Bar-Hillel, 'The present status of automatic translation of languages', in F.L. Alt (ed.), *Advances in computers*, vol. 1 (Academic Press, New York, 1969), pp. 92–163.

16. C. Shannon and J. McCarthy (eds), 'Automata studies', *Annals of Mathematics Studies, No. 34* (Princeton University Press, Princeton, NJ, 1956).

17. Comments by J. McCarthy, at an 'AISB Summer School on Expert Systems', held at Edinburgh University, July 1979.

18. The Dartmouth Conference is mentioned in several places, for example, M.L. Minsky, 'Artificial Intelligence' in *Information* (a Scientific American book) (Freeman, San Francisco, 1966), p. 194; and in P. McCorduck, *Machines who think* (Freeman, San Francisco, 1979), pp. 93–114.

19. A. Newell and H.A. Simon, 'The logic theory machine', *IRE Trans. on Info. Theory*, no. IT-2 (1956), pp. 61–79.

20. G.W. Ernst and A. Newell, GPS: *A case study in generality and problem solving* (Academic Press, New York, 1969).

21. J. McCarthy, 'Recursive functions of symbolic expressions and their computation by machine', part 1, *Comm. ACM*, vol. 3 (1960), pp. 184–95.

22. M.L. Minsky, 'Some methods of Artificial Intelligence and heuristic programming' in *Mechanisation of thought processes* (HMSO, London, 1969), pp. 3–28.

23. In addition, the 'Matthew Effect' would be operating: R.K. Merton, 'The Matthew Effect in science', *Science*, vol. 159 (3810) (5 Jan. 1968), pp. 56–63.

24. J. McCarthy, 'Toward a mathematical theory of computation' in *Proc. IFIP Congress 1962* (North Holland, Amsterdam, 1963).

25. M.L. Minsky (ed.), *Semantic information processing*, (MIT Press, Cambridge Mass., 1968).

26. M.L. Minsky, 'A framework for representing knowledge' in P. Winston (ed.), *The psychology of computer vision* (McGraw Hill, New York, 1975), pp. 211–77.

27. See, for example, M.L. Minsky and S. Papert (and staff), 'Proposal to ARPA for research on Artificial Intelligence at MIT 1971–1972', *MIT AI Lab. Memo.*, No. 245 (Oct. 1971).

28. A. Newell and H.A. Simon, *Human problem solving* (Prentice-Hall, Englewood Cliffs, N.J., 1972).

29. W.S. McCulloch and W. Pitts, 'A logical calculus of the ideas immanent in nervous activity', *Bull. Maths. Biophysics*, vol. 5 (1943) pp. 115–33.

30. Indeed, Simon received the 1978 Nobel Prize in Economics, and his 1947 book, *Administrative behaviour* (Macmillan, New York), was explicitly mentioned in the prize announcement.

31. As noted in McCorduck, *Machines who think*, p. 109ff.

32. J.M. Brady, 'Report on a visit to the U.S.', Essex University, 1975 (edited version of a report submitted to the Science Research Council, Sept. 1975), outlines the funding situation for Artificial Intelligence in the United States. McCorduck, *Machines who think*, p. 117ff., discusses the 'no strings attached' role of the Air Force funding in allowing the work of Simon and Newell to get started.

33. For attacks on the field see: M. Taube, *Computers and common sense: the myth of thinking machines* (Columbia University Press, New York, 1961); H.L. Dreyfus, *What computers can't do: the limits of Artificial Intelligence* (Harper and Row, New York, 1972); and J. Weizenbaum, *Computer power and human reason* (Freeman, San Francisco, 1976).

34. N. Elias, 'Scientific establishments' (fn. 12).

35. P. Armer, 'Attitudes toward intelligence machines' in E.A. Feigenbaum and J. Feldman (eds), *Computers and thought* (McGraw Hill, New York, 1963), pp. 389–405. This was also commented upon in interviews: R.A. Brooker, Essex University, 28/3/79; and A.M. Uttley, Sussex University, 21/3/79.

36. A.M. Turing, 'Intelligent machinery', (1947) in B. Meltzer and D. Michie (eds), *Machine intelligence 5* (Edinburgh University Press, Edinburgh, 1969), pp. 3–23; and A.M. Turing, 'Computing machinery and intelligence', *Mind*, vol. 59 (1950), pp. 433–60.

37. R.J.W. Craik, *The nature of explanation* (Cambridge University Press, Cambridge, 1952), p. 57.

38. W.R. Ashby, *Design for a brain* (Wiley, New York, 1952); and

An introduction to cybernetics (Methuen, London, 1965).

39. W.G. Walter, 'An imitation of life', *Scientific American*, vol. 182, no. 5 (1950), pp. 42–5; 'A machine that learns', *Scientific American*, vol. 185, no. 2 (1951), pp. 60–3.

40. F.H. George, *The brain as a computer* (Pergamon Press, Oxford, 1961).

41. The RATIO club was discussed in an interview: A.M. Uttley, Sussex University, 21/3/79; and is also discussed by McCorduck, *Machines who think*, p. 59.

42. The proceedings of this conference are published in *Mechanisation of thought processes* (HMSO, London, 1960).

43. J. McCarthy, 'Programs with common sense', ibid., pp. 75–84.

44. D. Michie, *On machine intelligence* (Edinburgh University Press, Edinburgh, 1974), pp. 37, 51, and 66–7; where he states: 'It was from my personal association with Turing during the war and early post-war years that I acquired my interest in the possibilities of using digital computers to simulate some of the higher mental functions that we call "thinking".'

45. This, and much of the following information pertaining to Michie's activities is derived from interviews with Michie (Edinburgh University, 29 and 31/8/78), backed up by other available sources.

46. For example, D. Michie, 'The effect of computers on the character of science', *University of Edinburgh Gazette*, vol. 34 (Oct. 1962), pp. 23–8; and 'The computer revoluton: where Britain lags behind', *The Scotsman*, 12/7/63.

47. *Computing science in 1964*, A Pilot Study of the State of University Based Research in the U.K., prepared for the Research Grants Committee by Dr. Donald Michie (The Science Research Council, London, 1965).

48. Interview: Lord Halsbury, London, 18/12/79.

49. Council for Scientific Policy/University Grants Committee, *A Report of a Joint Working Group on Computers for Research* (HMSO, Cmnd. 2883, London, 1966).

50. Interview: Lord Halsbury, London, 18/12/79.

51. Reconstructed from information in the archives of the Department of Artificial Intelligence, and the Machine Intelligence Research Unit, Edinburgh University, and from interviews.

52. J. McCarthy, 'Review of the Lighthill report', *Artificial Intelligence*, vol. 5 (1974), pp. 317–22.

53. R.M. Burstall and R.J. Popplestone, 'POP-2 Reference Manual' in E. Dale and D. Michie (eds), *Machine Intelligence*, vol. 2 (Oliver and Boyd, Edinburgh, 1968), pp. 207–46.

54. Interviews: B. Meltzer, Edinburgh University, 1/8/77 and 23/1/79.

55. N.L. Collins and D. Michie (eds), *Machine Intelligence*, vol. 1 (Oliver and Boyd, Edinburgh, 1967). There have been nine workshops in the series to date.

56. E.W. Elcock, 'Report of the SRC Computer Research Group', Aberdeen University, May 1970; and interviews: A.M. Murray, Aberdeen University, 9/11/78; P.M.D. Gray, Aberdeen University, 23/7/79; and J.M. Foster, R.R.E. Malvern, 19/6/79.

57. Information on the activities of Clowes derives from a letter: M.B. Clowes, 23/11/78; and interview: M.B. Clowes, Sussex University, 22/3/79.

58. Interview: B. Meltzer, Edinburgh University, 23/1/79.

59. Interview: P.J. Hayes, Essex University, 27/3/79.

60. These conferences, 'The International Joint Conferences on Artificial Intelligence', have been one of the main organs of communication in the area, and a major outlet for publications in the area.

61. R.L. Gregory, *Eye and brain* (Weidenfeld and Nicholson, London, 1966).

62. Information on the robot is derived from numerous documentary sources in the archives of the Department of Artificial Intelligence, and the Machine Intelligence Research Unit, and from various interviews.

63. Interview: J.A.M. Howe, Edinburgh University, 24/1/79.

64. As was evident from various interviews, including one with S. Michaelson, the director of the Computer Science Department at Edinburgh University, on 18/7/79.

65. Lord Swann commented on the new departures in an interview: London, 10/10/79. At a press conference in connection with an open day for industrialists, he is reported as commenting that Edinburgh was among the top two centres of research, despite having no costly 'big science' projects. *University of Edinburgh Bulletin*, vol. 8, no. 2 (Oct. 1971).

66. Interview: R.L. Gregory, Bristol University, 25/7/79.

67. Conversational Software Ltd., launched in 1970.

68. Interview: D. Michie, Edinburgh University, 31/8/78. He commented that at one stage some 20 sources were involved.

69. *Computing Science Review* (Science Research Council, London, 1972), p. 17.

70. Ibid., p. 19. Compare also D. Michie, 'Schools of thought about AI', *University of Edinburgh, School of Artificial Intelligence, Experimental Programming Report*, no. 32 (1973).

71. *Artificial Intelligence: a paper symposium* (The Science Research Council, London, 1973), p. i.

72. This review was carried out by a Special Committee, chaired by Prof. N. Feather, and set up by the University Court. It reported in 1973, and extracts were made available to me by C.H. Stewart, at that time Secretary to the University, in a letter, 18/7/78.

73. As was evident from interviews and also from the Departmental Newsletter.

74. J. McCarthy and P.J. Hayes, 'Some philosophical problems from the standpoint of Artificial Intelligence' in B. Meltzer and D. Michie (eds), *Machine Intelligence*, vol. 4 (Edinburgh University Press, Edinburgh, 1969).

75. G.D. Plotkin, 'Automatic methods of inductive inference', unpub. PhD diss., University of Edinburgh, 1971.

76. Interview: R.A. Kowalski, Imperial College, London, 15/6/79.

77. From the Departmental Newsletter.

78. From the Minutes of the Round Table, a sort of departmental board.

79. Interview: Lord Swann, London, 10/10/79.

80. Various interviews.

81. D. Michie, 'A six year project to develop an intelligent problem solving system' (modified version of a seven-year project proposal to the SRC, Edinburgh University, 1972).

82. P.A. Ambler *et al.*, 'A versatile computer controlled assembly system', in *Proc. Third International Joint Conference on Artificial Intelligence* (Stanford, 1973), pp. 298–307; and *Artificial Intelligence*, vol. 6 (1975), pp. 129–56.

83. Sir James Lighthill, 'Artificial Intelligence: a general survey' in *Artificial Intelligence* (SRC, London, 1973), p. 7.

84. As is evident in the *AISB Newsletter, SIGART* (The American Equivalent), and in numerous other places.

85. In the public press there were articles in *The Times Higher Education Supplement*, 1/6/73 and 8/6/73, in *Science*, vol. 180 (June 1973), pp. 1352–3, and in the *New Scientist*, 22/2/73 and 1/3/73. There was a public debate on 4 July, 1973, at the Royal Institution, Albemarle Street, London, under the chairmanship of Sir George Porter, between Sir James Lighthill and Professors R.L. Gregory, J. McCarthy, and D. Michie on Sir James's theme: 'The general purpose robot is a mirage'. This was televised by the BBC as one of its 'Controversy' series and broadcast on 30 August 1973. Later, videorecordings of the debate were taken to the United States as the 'Lighthill Tapes', where they were shown to packed audiences around the centres of Artificial Intelligence.

86. Lighthill, 'Artificial Intelligence: a general survey', p. 1 (fn. 83).

87. In *Proposed new initiatives in computing and computer applications* (Science Research Council, Swindon, March 1979), p. 8, there is the comment: 'The Panel has no doubt that the reluctance of the present community to take up the challenge (of industrial robots research) is due at least in part to the general discouragement of Artificial Intelligence which took place in this country several years ago and that it is now up to SRC to take steps to remedy the situation.'

88. Reported in *AISB*, vol. 20 (July 1975), and in *The Times Higher Education Supplement*, 14/3/75, p. 9.

89. It is impossible to give definite figures in brief for Artificial Intelligence funding because of the multiplicity of sources, and because of the difficulty in deciding what money went to specifically Artificial Intelligence research of the type discussed in this paper; the SRC, for instance have sometimes included under the category of machine intelligence work that is clearly along cybernetic lines.

90. Sources detailed in fn. 56.

91. Interview: R.A. Brooker, Essex University, 27/3/79.

92. Interview: D. Michie, Edinburgh University, 31/8/78.

93. D. Michie, 'Machine intelligence in the cycle shed', *New Scientist*, vol. 57 (22 Feb. 1973), pp. 422–3.

94. Interview: Lord Swann, London, 10/10/79.

95. E.W. Elcock and D. Michie (eds), *Machine Intelligence*, vol. 8, (Wiley, New York, 1977).

96. J.E. Hayes, D. Michie and L.I. Mikulich (eds), *Machine*

Intelligence, vol. 9 (Ellis Horwood, Chichester, 1979).

97. Michie organised the 'AISB Summer School on Expert Systems', held at Edinburgh University, July 1979.

98. A. Bundy *et al.*, *Artificial Intelligence: an introductory course* (Edinburgh University Press, Edinburgh, 1978).

99. T. Winograd, *Understanding natural language*, (Edinburgh University Press, Edinburgh, 1972).

100. Various interviews, and fn. 27.

101. J.A. Robinson, 'A machine oriented logic based on the resolution principle', *Journal ACM*, vol. 12, (1965), pp. 23–41.

102. Fleck, 'The structure and development of Artificial Intelligence', pp. 44–8 (fn. 3).

103. D. Warren, 'PROLOG on the DEC system-10', paper presented at the AISB Summer School on Expert Systems, Edinburgh University, July 1979, notes some applications. The position of theorem proving was also discussed in several interviews.

104. Interview: R.M. Burstall, Edinburgh University, 10/8/77.

105. Interview: R.A. Brooker, Essex University, 27/3/79.

106. Interviews: J.M. Brady, Essex University, 26 & 28/3/79; P.J. Hayes, Essex University, 27/3/79.

107. J.E. Doran, 'Knowledge representation for archaeological inference', in E.W. Elcock and D. Michie (eds); *Machine intelligence*, vol. 8 (Wiley, New York, 1977), pp. 433–54.

108. Interviews: B. Anderson, Essex University, 30/7/79; Y. Wilks, Essex University, 8/8/79.

109. Information on movements through Essex was derived from interviews with Brady (see fn. 106), backed up by other sources.

110. Information on developments at Sussex University was derived from various interviews and documentary sources, in particular the interview: N.S. Sutherland, Sussex University, 21/9/79.

111. For example, A.M. Uttley, 'Simulation studies of learning in an informon network', *Brain Research*, vol. 102 (1976), pp. 37–53.

112. M.B. Clowes, 'On seeing things', *Artificial Intelligence*, vol. 2 (1971), pp. 79–116.

113. This organisation is frequently commented on in the *Annual Reports*, Sussex University, and is discussed by the first Vice Chancellor, Asa Briggs, in his 'Drawing a new map of learning' in D. Daiches (ed.), *The idea of a new university: an experiment in Sussex* (Deutsch, London, 1964), pp. 60–80.

114. Letter: Lord Briggs, 28/11/79.

115. 'Working party on School of Cognitive Studies', University of Sussex, June 1970.

116. Interview: M.A. Boden, Sussex University, 19/9/79.

117. M.A. Boden, *Artificial intelligence and natural man* (Harvester Press, Hassocks, 1977).

118. Interview: A. Sloman, Sussex University, 20/3/79.

119. A. Sloman, *The computer revolution in philosophy* (Harvester Press, Hassocks, 1978).

120. Interview: P.N. Johnson-Laird, Sussex University, 20/9/79.

121. For instance, approximately one-quarter of the MSc computer

science courses outlined in *Graduate Studies 1974–75* (CRAC, Cambridge, 1974), mentioned Artificial Intelligence, and it is often mentioned as an acceptable interest in job advertisements.

122. J.M. Brady, 'A glimpse of the future of AI', text of a lecture delivered at 'Computing 79', Sydney, Australia, August 1979, p. 2.

123. B.C. Griffith and N.C. Mullins, 'Coherent social groups in scientific change', *Science*, vol. 177 (1972), pp. 959–64.

124. I do not want to argue a 'great man' view of history here. The situation can be interpreted in terms of the emergence of socially defined possibilities which in the event were exploited by Michie. If Michie had not been, development in Artificial Intelligence in Britain would still have taken place with other people stepping in to a greater extent. Perhaps what is required is a 'great opportunities' view of history, with the emphasis on the socially given possibilities rather than on the people who exploit them.

125. T. Shinn, 'Scientific disciplines and organisational specificity: the social and cognitive configuration of laboratory activities', in N. Elias, H. Martins and R. Whitley (eds), *Scientific establishments and hierarchies*, Sociology of the Sciences, vol. VI (Reidel, Dordrecht, 1982), pp. 239–64.

126. See for instance, P. Kraft, *Programmers and managers: the routinisation of computer programming in the United States* (Springer Verlag, New York, 1977).

127. M.J. Mulkay, in *The social process of innovation* (MacMillan, London, 1972), concludes: '. . . intellectual migration, whether into established networks or into virgin areas, will normally be led by mature researchers of known repute. These men use their eminence to attract funds and graduate students; and in various ways they try to use their existing knowledge and techniques as a point of departure for the construction of the new intellectual framework' (p. 54). The pattern of development of Artificial Intelligence in Britain certainly conforms with the first part of what Mulkay writes, but it is difficult to square it with the second part. It would appear that the extent of articulation of the paradigmatic structure in the United States preempted attempts by migrants in Britain to use their existing knowledge in the construction of new intellectual frameworks, and left them scope only for negotiation over the organisational aspects.

128. See N.S. Sutherland, 'Neuroscience versus cognitive science', *Trends in Neurosciences*, vol. 2, no. 8 (1979), pp. i–ii, which discusses some of the particulars of this dominance.

129. Lighthill, 'Artificial Intelligence: a general survey', p. 1 (fn. 83).

130. For example, Sir Geoffrey Allen (Chairman of the Science Research Council), in a talk in the Department of Liberal Studies in Science, the University of Manchester, 17/5/79.

131. C.E. Shannon, 'Automatic chess player', *Scientific American*, vol. 182, no. 2 (1950), pp. 48–51.

132. See fn. 33.

133. Turing, 'Intelligent machinery' (fn. 36).

134. Fleck, 'The structure and development of Artificial Intelli-

gence', pp. 138–9 (fn. 3).

135. Such an impression was given by those in Artificial Intelligence (for example, interview: A. Bundy, Edinburgh University, 17/7/79) and those outside, in the course of interviews.

136. Dreyfus, *What computers can't do*, is the classic example here (fn. 33).

137. Weizenbaum, *Computer power and human reason* (fn. 33).

138. Taube, *Computers and common sense* (fn. 33).

139. E.W. Dijkstra, 'Programming: from craft to scientific discipline', *Proc. International Computing Symposium, 1977* (Liège, Belgium, April 1977), pp. 23–30.

140. The Japanese success in catching up with and then overtaking the leading Western economies in shipbuilding, automobiles, consumer electronics, capital goods, and general manufacturing hardly needs further comment here, but it certainly added substance to the threat posed by the 5th Generation Project.

141. *Summer 1980 JIPDEC Report*, (Japan Information Processing Development Center, Tokyo, September 1980).

142. D'Agapeyeff was a founder of CAP Ltd., London, at one time one of the biggest software houses in Europe.

143. During a talk entitled 'Challenges for expert systems', given at the Expert Systems: 81 Conference, London, December 1981.

144. For instance: A. d'Agapeyeff, 'Report to the Alvey Directorate on a short survey of expert systems in UK business', *Alvey News*, Supplement to Issue No. 4 (April 1984).

145. See Figure 3.1 in section 3.5 of this chapter.

146. McCorduck wrote a history of AI in the US: *Machines who think* (fn. 18).

147. E.A. Feigenbaum and P. McCorduck, *The Fifth Generation: Artificial Intelligence and Japan's challenge to the world* (Michael Joseph, London, 1983), quote taken from the prolog.

148. Department of Industry, *A programme of advanced information technology*, The report to the Alvey Committee (HMSO, London, 1982).

149. Quoted in J. Alvey, 'UK response to the Fifth Generation', *Electronics and Power*, May (1983), pp. 387–9.

150. See: 'The Fifth Generation: taking stock', *Science*, 30 Nov. (1984), 1061–3.

151. Information derived from research in progress on an ESRC funded project: 'The effective management of available expertise in Artificial Intelligence', held by D. Edge and J. Fleck (database at Jan. 1987).

152. From the same source.

153. For instance, see the survey by d'Agapeyeff, 'Report to the Alvey Directorate' (fn. 144).

154. K. Cornwall-Jones, PhD research in progress, Science Policy Research Unit, University of Sussex, 1986.

155. For instance, A. Bundy et al. (eds), *A catalogue of Artificial Intelligence tools* (Science and Engineering Research Council, Swindon, Spring 1983).

156. As can be seen from the papers presented in the annual series of conferences 'Expert Systems: 81–86'.

157. See *Summer 1980 JIPDEC Report*, p. 15 (fn. 141).

158. M. de Mey, *The cognitive paradigm* (Reidel, Dordrecht, 1982).

159. Compare Descartes's use of the hydraulics metaphor.

160. For example, I. Aleksander, whose work on a pattern recognition chip was based more on a neural net tradition than on the list processing AI approach, has been successful in claiming this as bona fide AI research.

161. For a discussion of the various positions taken by protagonists in these debates, see: J. Fleck, 'Artificial Intelligence and industrial robots: an automatic end for utopian thought?' in E. Mendelsohn and H. Nowotny (eds), *Nineteen eighty-four: science between utopia and dystopia* (Reidel, Dordrecht, 1984), pp. 189–231.

162. For instance Searle certainly created a stir with his 1984 Reith Lecture series and his 'chinese box' critique of AI: J. Searle, *Minds, brains and science* (British Broadcasting Corporation, London, 1984).

163. See B.J. Rooney, 'A Survey of the Technical, Social, and Possible Impact Areas of Expert systems Technology', unpub. MSc diss., University of Aston, 1983.

164. Cornwall-Jones, research in progrss (fn. 154).

165. Seminar presentation by K. Guy, of Science Policy Research Unit, at Edinburgh University, 11 March 1986.

166. This educational effort is amply illustrated by the many training seminars, tutorials, and advanced courses on offer from the various bodies in the area.

4

Frames of Artificial Intelligence

Joop Schopman

As the industrial revolution concludes in bigger and better bombs, an intellectual revolution opens with bigger and better robots. The former revolution replaced muscles by engines and was limited by the law of the conservation of energy, or of mass-energy. The new revolution threatens us, the thinkers, with technological unemployment, for it will replace brains with machines limited by the law that entropy never decreases (McCulloch 1951: 42).

4.1 INTRODUCTION

The aim of this essay is to sketch a close-up of a crucial moment in the history of Artificial Intelligence (AI), the moment of its genesis in 1956. This came about as a result of the choices made by a group of people who were dissatisfied with the then-prevailing scientific way of studying human behaviour. They considered their approach as radically different, a revolution — the so-called 'cognitive revolution'. Here, an effort will be made to evaluate their choices.[1] It is contended that by investigating these processes of paradigm change, a better understanding of AI might be gained. In the first section, an exposition will be given of the investigative method used, SCOST — the 'Social construction of science and technology' — and reasons will also be given as to why this method has been chosen. In the second section, a description will be given of the period before 1956. This will then be followed by a section on the nature of AI as conceived by its founders. As this is a better known phenomenon, it will be dealt with less extensively than the pre-1956 period. In the fourth section attention will be given to reasons for the transition of the pre-1956 period into the AI era. In the final section an evaluation will be made of that transition: it will be argued that the founders of AI have been overstating the originality of their approach, and that it would be more accurate to describe it as a (too hasty) step in an ongoing process, and certainly not as a revolution in the Kuhnian sense.

4.2 METHODOLOGICAL POSITION

Several methods are available for the study of the development of science and technology. Most commonly, one assumes that science and technology develop according to their own internal structure and logic. On this view, the interplay between theory, experiment, and the object of study can fully explain the actual development. External factors, such as funding, are important because they can influence the speed or location of developments. However, they have nothing to contribute to the internal course itself. Behind this approach lies the conviction that the acquisition of knowledge is a process in itself, i.e. is independent of humans. Of course, humans are needed for something to be known in the first place. This can be exemplified by the exploration of new territory such as Antarctica: we have to go there to find out what it is like, but our knowledge of that area is dictated by what exists in Antarctica and is independent of who is looking at it. Eskimos, Japanese or Dutch should all come to the same conclusions if they do their job properly. Their language will differ, but the content will be identical. This is the most conventional view, supported not only by historians and philosophers of science but also, perhaps more so, by scientists themselves. This way of describing scientific and technological developments is called *internalistic*. Only factors within science and technology can contribute to its growth. Those internal factors have to be understood restrictively — that is, no factors are allowed which belong to scientists and technicians as individual or social beings.

In reaction to this 'inhuman' way of understanding knowledge, some sociologists of science have formulated a radically different strategy. In their opinion, all knowledge is socially constructed: it is not the outcome of a process of discovering, of finding out how things are, but the result of an interaction between the participants of the knowledge-creating process. Thus, all scientific and technological knowledge can be understood as the outcome of a social process — or, put differently, an internal logic of scientific development does not exist. The structure of the development coincides with the structure of human interactions. One could call this approach the *externalistic* view of science. Where the first approach is characterised by 'truth' (correspondence with reality), this one is determined by 'power' (force of rhetoric).

As one might expect there are also middle positions. One of these can be called a mixed position because it states that during the development of a discipline there are periods determined purely by their internal logic, and other periods open to external influence. This position is taken by the 'Starnbergers'.[2]

Another position is that of Ludwik Fleck and his approach has been taken as the methodological starting point for this chapter. Fleck refuses to separate external and internal determinants of scientific development which he considers to be shaped within the 'style of thought' (*Denkstil*) of a group (*Denkkollektiv*). The creation of thought happens 'through' two poles: the social community and physical reality. Both determine the style of thought although the emphasis in his writings is more on social determinants. His ideas, though published in 1935, remained completely unnoticed. The work of Thomas Kuhn has made some of them known to a larger audience and in particular Kuhn's term 'paradigm' can be compared with Fleck's notion of 'style of thought'. However, the concept of a paradigm has provoked a great deal of criticism which, in a way, has forced Kuhn to retreat by filling out the term in a mainly sociological way.[3] However, for the purpose of understanding scientific practices, not only visible social structures are needed but all underlying 'layers' as well, including social-cultural traditions and philosophical ideas.[4] It is the coherence of these layers which the original concept of a paradigm was meant to emphasise.

To bring this point into effect and to avoid confusion of the actual concept, we will borrow another term from recent literature — namely, *frame* — which is a concept particularly used by those researching the social construction of technology (SCOT).[5] A frame consists of a whole conglomerate based on a

167

common worldview as well as disciplinary practices; these create the background for interactions so that opinions can be shared. These interactions are often described in terms of 'networks' (a multidimensional construct made of very different elements), an interwovenness of people, ideas, power, institutions, etc., which shapes a part of our (scientific) social reality. One rightly speaks about bicycles and ultracentrifuges as social constructs. It applies even more to less tangible things such as 'intelligence', let alone AI. A complicated interaction of a variety of actors shaped the genesis of AI. To take them into account the existing SCOT programme will be extended to science (SCOST). There is no pretence of presenting a complete picture here; only the most important actors will be dealt with. Thus, to study AI as a frame means that one has to take seriously its social component, its constructiveness. Simultaneously, however, the 'reality' part should not be forgotten: these two aspects characterise the concept of 'style of thought' as formulated by Fleck.

4.2.1 The 1956 turning point

It may seem surprising to see 1956 mentioned as the crucial date. An historic change can never be dated exactly: even when there is general agreement about the occurrence of a change, there is no reason to believe that it can be fixed at an exact date. Changes take time, even in the case of AI. In retrospect, however, there is a strong feeling that 1956 was the turning point; or better, the starting point. The Dartmouth Conference of 1956 is generally seen as *the* important event which put AI on the map, and accordingly, attention will be given to what made this conference so crucial. When the periods before and after 1956 are discussed it must be kept in mind that 1956 is of course an artificial divide. In fact it stands for a conceptual divide because it is seen as such by the AI community; but research that fits the AI-frame may in fact have been carried out before or after 1956.

4.3 THE PRE-1956 FRAME

The Dartmouth Conference initiated AI and marks the beginning of a new frame. Hence it may seem strange to call the period before 1956 a frame as well, but there are good reasons for doing so. From the middle of the 1930s until 1956 there were several disciplinary developments which interacted so that, in the SCOST terminology, a frame emerged within which a stabilisation process became possible between different actors.[6] Actors had formed their opinions due to developments in their own field and what they had experienced as parallel developments in neighbouring areas. Around 1943 a kind of common point of view developed out of their interaction — the pre-1956 frame. Within it the opinions of the individual participants contributed to common ideas — or, put differently, within the frame, points of view were the outcome of a stabilisation process, i.e. an interactive process in which individual opinions and social positions determined the outcome. The implication of this is that any final opinion does not need to coincide with any one person's private opinion. On the other hand, opinions are temporal because they are the outcome of a stabilisation process. A change of position of one of the participants or a change in composition of players will change the outcome, i.e. the 'shared' opinion. Obviously, during the development of AI both events occurred.

Clearly, developments in logic at the turn of the century set the stage for those trends which later dominated the pre-1956 and AI frames. In fact, pre-1956 developments in logic had a huge impact on the education and thinking of researchers in other fields, such as philosophy, neurophysiology, and mathematics. Logic appeared to be the ideal tool for analysing human thinking. In 1936 Alan Turing had published an article in which he demonstrated that what can be calculated 'mechanically' can be calculated on a simple 0–1 code manipulating device. In 1932 the Nobel Prize was awarded for research in neurophysiology which had concluded that the human nervous system could be described as a complex network of neurons which are, in principle, yes-no devices. Thus, the human nervous system could, ideally, be described as a logic system. These developments led to the opinion that human behaviour could be understood as the outcome of 0–1 processes. This conviction was confirmed by the construction of computers in the 1940s and 1950s, and the digital computer in particular was seen as performing structurally in much the same way as the neural system operates. Thus, one could say that in the pre-1956 period the digital computer had become *the* metaphor for human intelligent behaviour. I will refer to this way of comparing human and computer performances as *the hardware metaphor*.

After this brief indication of the pre-AI developments, the contribution of some of the main actors will be discussed in more detail. We will start with the contribution of McCulloch. His work predates that of the other participants to be described here and was a source of inspiration to such mathematicians as Wiener and von Neumann, who will be discussed subsequently. (All this might give the impression that only the USA was in the game, but the UK's contribution was at least as important, although of a somewhat different character.) Finally, attention will be given to Boring, who was a representative of the great absentees: the psychologists. Their discipline, at least the American part of it, was so encapsulated in its efforts to operate scientifically, that most of them did not have any real contribution of their own to make. They merely absorbed the hardware ideas referred to above. Something different was needed to get them moving.

4.3.1 Warren S. McCulloch

As mentioned earlier, developments in neurophysiology contributed to the hardware metaphor. We will not go into the neurology itself as a frame, but will take just one outspoken representative who was very influential in the interaction process which shaped the hardware metaphor. Warren S. McCulloch (1898–1969) connected physiological results with modern logic. He argued that the all-or-none character ascribed to neural activity could be represented in the binary system of logic, or in propositions. This made it possible to conceive the response of any neuron 'as factually equivalent to a propositon which proposed its adequate stimulus' (McCulloch 1965/1943: 21). Because of the fact that physiological relations between neural activities corresponded with relations between propositions, the behaviour of complex nets could be represented by the symbolic logic of propositions (Figure 4.1). However, facilitation and learning, i.e. temporary and lasting effects of stimulations, turned out to be problematic.

Figure 4.1: Complex nets and symbolic logic

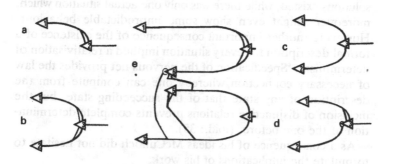

EXPRESSION FOR THE FIGURES

In the figure the neuron c_i is always marked with the numeral i upon the body of the cell, and the corresponding action is denoted by 'N' with i as subscript

Figure 4.1a $N_2(t) . \equiv . N_1(t - 1)$

Figure 4.1b $N_3(t) . \equiv . N_1(t - 1) \vee N_2(t - 1)$

Figure 4.1c $N_3(t) . \equiv . N_1(t - 1) . N_2(t - 1)$

Figure 4.1d $N_3(t) . \equiv . N_1(t - 1) . \sim N_2(t - 1)$

Figure 4.1e $N_3(t) . \equiv . N_1(t - 1) . \vee . N_2(t - 3) . \sim N_2(t - 2)$

$N_4(t) . \equiv . N_1(t - 2) . N_2(t - 1)$

Source: McCulloch 1965/1943: 36s

In 1943 McCulloch and the mathematician Walter Pitts solved these problems for systems with or without 'order' (i.e. feedback/recursive function). For every (altered) net there was a corresponding logical expression and (under certain conditions) vice versa.[7] A system which could reduce the difference between the input of a net and activity in the net exhibited purposive behaviour.

Thus both the formal and the final aspects of that activity which we are wont to call *mental* are rigorously deducible from present neurophysiology. (. . .) Thus in psychology, introspective, behaviouristic or physiological nature, the fundamental relations are those of two-valued logic (McCulloch 1965/1943: 38).

They emphasised the fact that formal solutions can never replace real ones: for each situation, many equivalent formal solutions existed, while there was only one actual situation which, moreover, might even show some unpredictable behaviour. However, another important consequence of the existence of a formal description for every situation implied a relativisation of determinism: 'Specification of the nervous net provides the law of necessary connection whereby one can compute from the description of any state that of the succeeding state, but the inclusion of disjunctive relations prevents complete determination of the one before, (ibid.: 35).

As a consequence of his ideas McCulloch did not hesitate to formulate the implications of his work:

> we have constructed automata which, like us, can compute any computable number, can formulate clear ideas, and, by inverse feedback, have purposes of their own, built into them as ours are born in us, (. . .) As yet we have not made them capable of multiplying their kind. That would be for us the final mistake (McCulloch 1965/1948: 155).

At the Hixon Symposium (1948) he expressed his ideas more systematically. Only two kinds of things existed: 'body' or 'stuff and process', which are observable phenomena; and 'mind' or 'idea and purpose', which belong to the realm of discourse and as such are not observable. No physics is possible without objects, and in their turn no objects exist without people. Humans detect regularities and invariances using logic and mathematics, but the regularities themselves are neither stuff nor process — that is, as 'ideas', they don't belong to physics. How then could their existence be explained? The only way to do this seemed to be in neurophysiological terms, but then a better way appeared:

> Now Robert Wiener has proposed that information is orderliness and suggests that we measure it by negative entropy (. . .) Let us, for this argument, accept his suggestion. Ideas are then to be construed as information. Sensation becomes entropic coupling between us and the physical world (. . .) The attempt to quantify the information leads to a search for an appropriate unit (McCulloch 1951: 431).

That left the choice between digital and analog devices. The nervous system was *par excellence* a logical machine. The sense organs and effectors were analog but they reported via discrete pulses. Thus, somewhere in the conversion from analog signals to digital ones ideas were born.[8] On this account, digital systems seemed to be basic: Wiener proposed to define the unit of information for bivalent systems as the decision as to *which* state it shall occupy. By calculation Pitts and McCulloch demonstrated that the capacity of neural systems is many times larger than is needed for the information which they receive. This enables an organism to build a statistically reliable signal which secures a high probability that what passes through the nervous system does correspond to something in the world.[9] In this, organisms surpass all machines.

What was new here was the combination of Wiener's quantification of information with his previous ideas based on the bivalency of neural activities. Any nervous impulse could be treated as an atomic event, but also as a signal: 'Thus nervous impulses are atomic signals, or atomic propositions on the move. To them the calculus of propositions applies provided each is subscripted for the time of its occurrence and implication given a domain only in the past (McCulloch 1951: 47). In nets devoid of circles, the origin of a signal which entered a system could be traced back, but by introducing circles its context remained, as an 'eternal idea', as long as the signal was contained in the system (even though the circumstances of its origin would have disappeared). However,

> there are other closed paths important in the origin of ideas, circuits which have 'negative feedback'. (. . .) They are, as we say, error-operated. The state toward which they return the system is the goal, or aim, or end *in and of* the operation. This is what is meant by function. On these circuits Cannon founded his theory of homeostasis, and Rosenblueth and Wiener their theory of teleological mechanisms (ibid.: 47–8).

Thus, here he had integrated Wiener's work on information and feedback. The rest fitted nicely within the logical positivist tradition, as for example formulated by Carnap, wherein it was contended that logic related sense inputs to basic concepts. In Carnap's case it was not evident how the relation between sense impressions came about. McCulloch used neurophysiological data; sense impressions aroused neural activities which were logic. He also went along with Carnap's notion that all higher concepts were logical constructs (calculus, eventually with in-built feedback) of atomic ones. Even further, he reasoned that when the output of one calculator of invariants is made the input to another, there is an idea of ideas, 'what Spinoza calls consciousness, and thus get far away from sensation. But our most remote abstractions are all ultimately reducible to primitive atomic propositions and the calculus of the lowest level' (ibid.: 54).[10] McCulloch stated his belief boldly but he was also well aware of the problems: for the function of even the simplest structure of the brain was unknown.

He finished his contribution 'Why is the mind in the head' with the remark that the head is the place where most possible connections could be formed.

> Each new connection serves to set the stage for others yet to come and better fitted to adapt us to the world, for through the cortex pass the greatest inverse feedbacks whose function is the purposive life of the human intellect. The joy of creating ideals, new and eternal, in and of a world, old and temporal, robots have it not. For this my Mother bore me (McCulloch 1951: 56–7).

4.3.2 Norbert Wiener and John von Neumann

Mathematicians have played a most important role within the frame of the hardware metaphor. One of the prominent ones was Norbert Wiener (1894–1964) who after his thesis (1912) at Harvard in the philosophy of mathematics, spent a year at Cambridge (UK) studying logic and mathematics with Hardy and Russell. It was the latter who gave him his sense of the physical world and this set the trend for his future life.

From 1919 onwards, Wiener worked at MIT where he continued this preference for application to the physical world. By this time he had become influenced by Walter Gibbs's work on statistical mechanics and he started to consider the general-isation of the concept of probability. Its application to the problem of turbulence proved too difficult; a better choice turned out to be the problem of the Brownian motion. This theory also proved applicable to the so-called shot effect, the vacuum tube or conductor noise, and thus important for electrical engineers some 20 years later (e.g. radar).

Another impulse came from the department of electrical engineering where Professor Dugald C. Jackson was head. Thus, Wiener's mathematical thinking also became linked to the then-current technical developments. The branch of 'communication' engineering had still some way to go.

Wireless had been an established art for twenty years, but for the most part this was wireless in the limited sense in which Marconi had conceived it. Broadcasting was yet to come on a national scale. (. . .) The electronic valve had indeed arrived. The telephone, indeed, was trimphant everywhere, and was extending around the world, the tentacles of a tight communication net (Wiener 1966/1948: 72–3).

The work done thus far needed extension in order to be applicable to this new demand.

'I had to study harmonic analysis on an extremely general basis, and I found out that Heaviside's work could be translated word for word into the language of this generalized harmonic analysis. In all this there was an interplay between what I was doing on the Heaviside theory and what I had done on the Brownian motion (ibid.: 77–8).

However, the mathematics he had developed proved to be extendable to continuous phenomena as well.

I found that it was possible to generate continuous spectra by means of the Brownian motion or the shot effect and that if a shot-effect generator were allowed to feed into a circuit that could vibrate, the output would be of that continuous character. In other words, I already began to detect a statistical element in the theory of continuous spectrum, and through that, in communication theory (ibid.: 79).

In the meantime Vannevar Bush was working at MIT on high speed computing machines for solving problems of differential equations in connection with electrical circuits. He used measured quantities and thus analog techniques. In 1932–3 Wiener's friend Manuel Vallarta introduced him to the Mexican physiologist Arturo Rosenblueth with whom he shared an enthusiasm for scientific methodology.

At the outbreak of the Second World War Wiener considered the development of a digital computer as a possible contribution to the US war effort. He wanted to change the push-hole card machines made by IBM into ones which used electric sparks that could be magnetically recorded: 'Bush recognised that there were possibilities in my idea, but he considered them too far in the future to have any relevance to World War II' (ibid.: 239).[11] The refusal forced Wiener to move into a different area — namely, a control system for anti-aircraft guns. After some promising experimental runs with Bush's differential analyser, this project became a classified one to which Julian Bigelow became assigned. The problem that they faced meant that they had to simulate the flight of an aeroplane as well as the movement of an anti-aircraft gunner. Actions of the latter could be mechanised using the developments or radar but this was inapplicable to the former. To solve the problem mathematically it was

> necessary to assimilate the different parts of the system to a single basis, either human or mechanical. Since our understanding of the mechanical elements of gun pointing appeared to us to be far ahead of our psychological understanding, we chose to try to find a mechanical analogue of the gun point and the airplane pilot (ibid: 251–2).

This was done by negative feedback, although excessive feedback produced undesirable oscillations. In the eyes of Bigelow and Wiener this 'pathologic' behaviour could be compared with human pathologic conduct. They put their question to Rosenblueth who answered

> that such pathological conditions are well known, and are termed intention tremors (. . .) Thus, our suspicion that feedback plays a large role in human control were confirmed by the well-established fact that the pathology of feedback bears a close resemblance to a recognised form of the pathology of orderly and organised human behaviour (ibid.: 253–4).

In 1943 they wrote an article together which opened with the remark that the behaviouristic efforts to correlate inputs and outputs of an object were inadequate because they disregarded its specific structure and internal organisation. Alternatively, the functional approach focused on such structures and considered their relationships to the surrounding environment as relatively incidental. The authors classified different forms of behaviour in order to focus attention on those machines which were intrinsically purposeful (so-called servo-mechanisms) with negative feedback (teleological mechanisms), i.e. devices in which the output signal is fed back to control the input.

Their conclusion was that there was a considerable overlap between the behaviour of machines and organisms. They were so impressed by the success of this analogy that they raised the question of how far it went.

> While the behaviouristic analysis of machines and living organisms is largely uniform, their functional study reveals deep differences. Structurally, organisms are mainly colloidal, and include prominently protein molecules, large, complex and anisotropic; machines are chiefly metallic and include mainly simple molecules. (. . .) In future years, as the knowledge of colloids and proteins increases, future engineers may attempt the design of robots not only with a behaviour, but also with a structure similar to that of a mammal (Resenblueth, Wiener, Bigelow 1943: 22–3).

Given projected developments in computing and AI, and in particular the idea of bionic components, they displayed an extreme farsightedness.

As it happened, that same year McCulloch and Pitts published the paper which brought the comparison between the nervous system and a logic system to Wiener's attention. Coloured by the appearance of digital computers, he stated in retrospect that he had always been interested in computing machines but not so much in those made of metal and glass. The all-or-none law of the nervous system was not the only reason he felt compelled to regard the nervous system in much the same light as a computing machine.[12]

> The nerve fiber is a logical machine in which a later decision is made on the basis of the outcome of a number of earlier

decisions. This is essentially the mode of operation of an element in a computing machine. Besides this fundamental resemblance, we have auxiliary resemblances pertaining to such phenomena as memory, learning and the like (Wiener 1966/1967: 291).

However, he also placed a strong restriction on the use of logic. *'All logic is limited by the limitations of the human mind when it is engaged in that activity known as logical thinking.'* (Wiener 1962/1948: 125).

In 1944 Wiener gathered a group of neurophysiologists, communication engineers and computing machine men at Princeton. He considered this meeting as the birth place of the new science of cybernetics. The question concerning the domains to which this theory could be extended formed a topic at meetings which have been organised since 1946 by the Josiah Macy Foundation, the nucleus of which

> has been the group that had assembled in Princeton in 1944, but Drs. McCulloch and Fremont-Smith have rightly seen the psychological and sociological implications of the subject, and have co-opted into the group a number of leading psychologists, sociologists and anthropologists (ibid.: 18).

However, Wiener himself did not want to extend his work to social disciplines although Gregory Bateson and Margaret Mead urged him to do so. He felt, however, that this domain was unsuitable:

> For a good statistic of society, we need long runs *under essentially constant conditions*, just as for a good resolution of light we need a lens with a large aperture. (. . .) Thus the human sciences are very poor testing-grounds for a new mathematical technique (ibid.: 25).

He described his ideas in a book whose title summarises his aims: *Cybernetics, or control and communication in the animal and the machine* (1948). In summary, then, the importance of Wiener's work was not only that of imparting impetus to interdisciplinary research, but also that of stressing the autonomy of organisms with regard to their interaction with the environment. In addition, he realised that the role logic might play was not as simple as McCulloch had suggested.

Another mathematical prodigy, John von Neumann (1903–57), who had always been interested in physical problems, was influenced in his thinking by the work of McCulloch and Pitts. Moreover, during the Second World War, his attention had been drawn by Goldstine to the computing devices then under construction. This led to his interest in the mathematical side of automata.

> Their role in mathematics presents an interesting counterpart to certain functional aspects of organisation in nature. (. . .) Some regularities which we observe in the organisation of the (natural organism) may be quite instructive in our thinking and planning of the (artificial automata); and conversely, a good deal of our experiences and difficulties with our artificial automata can be to some extent projected on our interpretations of natural organisms (v. Neumann 1951: 1–2).

The complexity of organisms led, according to him, to a division of labour: physiology studied the parts, while mathematics concentrated on the organisation into a whole. The parts were to be treated at automatisms, the inner structure of which need not be disclosed.[13]

Artificial automata (computing machines) could be divided into digital and analog machines, as with organisms. Analog devices are based on the principle that numbers are represented by certain physical quantities, but the inherent noise in such devices limits their accuracy.[14] Von Neumann realised that there were objections to the idea of the neuron as an exact digital organ. This was relevant because a 'fully developed nervous impulse, to which an all-or-none character can be attributed, is not an elementary phenomenon, but is highly complex' (v. Neumann 1951: 10). However, that did not make the comparison less attractive because the digital computer was confronted with similar problems. Its main component at the time, the vacuum tube, was also a complex phenomenon, a 'complicated analogy mechanism'. Therefore, he defined something as an all-or-none organ when it so functions under normal operating conditions.

I realize that this definition brings in rather undesirable

criteria of 'property' of context, of 'appearance' and 'intention'. I do not see, however, how we can avoid using them, and how we can forego counting on the employment of common sense in their application (ibid.: 11).

However, as a working hypothesis, neurons and vacuum tubes could be taken as 'relay organs'.

At that moment (1948) progress was hampered by the technical limitations of the computers then available, as well as by the lack of a mathematically logical theory of automata. The existing formal logic dealt with rigid, all-or-none concepts. However, computers were real-time machines made of fallible components, so logic had to take probabilities and errors into account. There were, however, even more fundamental problems: in a discussion of the McCulloch-Pitts theory — which asserted that any functioning of an organism which could be defined logically and unambiguously in a finite number of words could also be realised by a formal neural network[15] — von Neumann questioned whether it could be done within the physical limitations of an organism, or whether every existing mode of behaviour could really be put completely and unambiguously into words.[16] The first question belonged to physiology; the second to the problem that

> there is no doubt that any special phase of any conceivable form of behaviour can be described 'completely and unambiguously' in words. This description may be lengthy, but it is always possible. To deny this would amount to adhering to a form of logical mysticism (v. Neumann 1951: 23).

Drawing on the example of a comparison of two geometrical figures he doubted whether a complete description could be possible. The order of complexity was out of all proportion and without any indication that logic could handle it.

> It is, therefore, not at all unlikely that it is futile to look for a precise logical concept, that is, for a precise verbal description, of 'visual analogy'. It is possible that the connection pattern of the visual brain itself is the simplest logical expression or definition of this principle. (. . .) All this does not alter my belief that a new, essentially logical, theory is called for in order to understand high-complication automata

and, in particular, the central nervous system. It may be, however, that in this process logic will have to undergo a pseudomorphosis to neurology to a much greater extent than the reverse (ibid.: 24).

Thus, what appeared to have been a quantitative problem looked as though it might turn out to be a qualitative one.

Von Neumann stressed again the differences between computers and brains in his unfinished book 'The computer and the brain'. After remarking that the intention of a machine was the intention of the user impressed upon it,[17] he stated their typical differences: recovery time, size, and energy consumption. All this led him to conclude that

parallel and serial operation are not unrestrictedly substitutable for each other. (. . .) More specifically, not everything serial can be immediately parallelled — certain operations can only be performed after certain others, and not simultaneously with them. (. . .) In such a case, the transition from a serial scheme to a parallel one may be impossible, or it may be possible but only concurrently with a change in the logical approach and organisation of the procedure. Conversely, the desire to serialize a parallel procedure may impose new requirements on the automaton (v. Neumann 1958: 51).

In his opinion, taking the neuron as a digital device was an idealisation and a simplification: their construction, interrelation, threshold and summation time made even an individual neuron 'a much more complicated mechanism than the dogmatic description in terms of stimulus-response, following the simple patterns of elementary logical operations, could express' (ibid.: 56). Furthermore, the reaction of receptors was not of a single neuron but of more complicated neuron systems. All this forced him to suggest that the nervous system could not be considered merely as a digital system; it contained analog or even mixed systems.

At the end he philosophised about the logical structure of the brain. Every automaton constructed for human use had a numerical and logical part. When we look at the nervous system as a computing machine, the question arises as to what accuracy should be expected of its numerical part. According to von Neumann, to perform the task the brain carries out, it must

181

have an extremely high accuracy. Thus, as a digital machine the brain could not succeed. Such accuracy would have required many successive steps, i.e. high error probability and high time consumption. Actually, the precision of levels of brain processes cannot be better than two or three decimal places. 'This fact must be emphasized again and again because no known computing machine can operate reliably and significantly on such a low precision level' (ibid.: 78). However, that did not make the brain an unreliable system. It could do so well because essentially it operated statistically on the outputs of parallel systems.

The last remark von Neumann made concerned language. For him logic and mathematics were accidental forms of expression. Other systems existed apart from the ones we were used to. 'Thus logic and mathematic in the central nervous system, when viewed as languages, must structurally be essentially different from those languages to which our common experience refers,' (ibid.: 82). This remark was repeated in his final lines: 'However, the above remarks about reliability and logical and arithmetical depth prove that whatever the system is, it cannot fail to differ considerably from what we consciously and explicitly consider as mathematics' (ibid.).

In the nine years that passed between his contribution to the Hixon Symposium (1948) and his last book, von Neumann did not become any less hesitant about the computer metaphor. The brain was certainly not a digital machine; so if it was to be compared with a machine, it must be a machine which functioned quite differently from any of the known devices.

4.3.3 Donald M. MacKay and Walter Ross Ashby

Although the European part of the AI equation does not seem to have played such a dominant role as that of the USA, similar trends were in fact evident in the UK. These were not restricted to the work of Alan Turing, who did such important theoretical and practical work on computers and their possibilities. Indeed, several people were interested in the subject and most met each other at the Ratio Club which had started on 14 September, 1949.

One of them was the physicist Donald M. McKay who was interested in the question of whether digital computing machines could show behaviour of the type which we normally regard as characteristic of the human mind. Thus far, most machines seemed to have avoided this resemblance: 'originality, independence of opinion, and the display of preferences and prejudices are not favoured by the mathematician in his computing tool' (MacKay 1951: 105). To address the question he reasoned in three steps: firstly, a machine could display the properties of an *organism*, e.g. goal-directed behaviour (purpose, control). A feedback system not only reacted to outside physical variables but also to internal ones (he referred to Ashby's 'homeostat'). Moreover, an artefact could reason and interact with humans in a logical dialogue, as for example in chess playing. As a result of gained experience an artifact could change its own pattern of behaviour and 'develop quite complex and superficially unpredictable characteristics capable of rational description in purposive terms' (ibid.: 108). Thus, from this he concluded that:

any pattern of observable behaviour which can be specified in terms of unique and precisely-definable reactions to precisely-definable situations can in principle be imitated mechanically. (. . .) In principle such artefacts could merit the title of organisms (ibid.: 108–9).

Secondly, he considered whether an artefact could simulate *mindlike* behaviour, e.g. originality, independence of opinion, even prejudice. In human discourse it is necessary to work with 'partial truth' or probability. 'This implies that the *statistical structure* of its behaviour pattern shall be that logically justified by the statistics of the data if it is to be described as reasonable and unprejudiced,' (ibid.: 109). That could be done via threshold control but random changes would cause unpredictable behaviour. 'Such an element would behave as if *uncertain* of the truth of the proposition represented by the stimulating signal' (ibid.: 110). However, the disturbance did not have to be random; a functional disturbance would have the same effect. This would have given a stochastic reasoning mechanism and the network guiding the statistical computations would in many cases be more important than the input itself. The exhibited behaviour was still sensical, because although its performances were not predictable individually, they were statistically. Within this view, mindlike behaviour could be simulated.

Prejudices, preference, and other 'emotional' effects can obviously be shown if transition-probabilities are linked appropriately to the 'causes' of the prejudices. *Weighing* of evidence can be represented by subliminal manipulation of probability-amplitudes. *Originality* of a kind can be constantly in evidence because of the lack of a deterministic link between input and output (ibid.: 111).

The system could also *learn* from past failure and success. MacKay built such devices wherein information was not stored locally but was immediately available. From all this he could conclude more stringently: 'Any pattern of observable behaviour which can be defined *statistically*, in terms of *probable* reactions to given situations, can in principle be shown to be an artefact' (ibid.: 112).

The third and final step concerned *abstraction* and hypothesis formation; how, for example, could a triangle be recognised? One possibility was to use the template-fitting method which McCulloch and Wiener had proposed, but MacKay himself suggested another approach — one which could be termed operational. This was based on the acts needed to replicate the outlines of an object: '*The elementary acts of replication define the basic vocabulary in terms of which the artefact describes its own experience*' (ibid.: 114). By taking as elementary symbols the elementary acts of replication there was no problem of identification. Complex concepts were represented by complexes of symbolic (internal) acts of representation. The complex of responses to something became the meaning of its name. However, because replication was never perfect, the meaning of a symbol could be more general than the specifications of a unique set of 'eigensymbols', or a probability distribution over sets of them. When the response-control mechanism was changed by an artifact, new replications had to be learned through statistical convergence. This constituted a first level of abstraction.

The upshot of this reasoning was the view that it would be incorrect to separate system and environment: mindlike and physical features each had their own role to play, and they belonged together as complements. In considering the answer to his original question, MacKay concluded that in principle no distinction could be found between the observable behaviour of a human brain and the behaviour possible in a suitably designed artefact.

As one of the contributors to the Dartmouth conference, MacKay stated that an intelligent robot needed some internal physical representation. The main question was about whether this was a given, limited to the cognitive categories of the designer, or whether an automaton could develop and learn new concepts? In the former case — the position taken by McCulloch and Pitts — the incoming signal was first coded so that it became invariant for transformations. In the latter there was no standard coding process. The system tried to match the input with actual content and to symbolise the features necessitating modifications. This symbolisation process was not input-oriented, i.e. a kind of filtrate of the input, but outwardly directed to a response. As a result of experience, an error-sensitive automaton obtained an optimal statistical structure. When a particular input had become adapted and was symbolised by the corresponding organising route, then it had entered 'the conceptual vocabulary of the automaton'.[18] This type of device functioned optimally when information was redundant. For real time processors this was very important. It was helped by the complexity of the error signal. The larger the information was, the shorter the response time. Again, it was not the world in isolation that was represented, but 'the nature and extent of the mismatch between the automaton and the world' (MacKay 1956: 248) as soon as a response route had been found.

MacKay was not afraid of bold terminology: an automaton could evolve a conceptual vocabulary of the same order as a human one; or, if the activities of its organising routes could be used as data there was 'the possibility of symbolic activity representing in effect *metalinguistic* concepts,' (ibid.: 249) 'Operationally it is as if the automaton is conscious only of the features of the world symbolized by its own internal matching-responses, and only indirectly of the error-signals evoking them' (ibid.: 250). He ended with the remark: 'Any resemblance between such an automaton as described and the human brain, is scarcely coincidental, but is logically inadmissible as evidence' (ibid.). His approach resembled those already mentioned but it was also entirely different. Nowhere in his system was a logical or digital step needed. MacKay actually built simple analog devices which could learn from experience. Still better known was Walter Grey's turtle and Ashby's 'homeostat'.

Walter Ross Ashby's work involved a serious effort to construct a machine which could adapt its behaviour. He worked on this problem in order to understand how the brain itself could produce behaviour: he wanted to show that this was due to a method used by the brain which was then available to machines. Of the two types of neural activities, one was reflex, inborn, genetically determined in detail, and produced — we might assume — by the physical/chemical structure of neurons; the other type was learned and this normally meant an adaptive change for the better. In simple cases, two related reactions might be seen as independent events caused by separate mechanisms. However, in complex cases this approach was insufficient to explain the reactions.

In most cases the 'correct' and the 'incorrect' neural activities are alike composed of excitations, of inhibitions, and of other processes each of which is physiological in itself, but whose correctness is determined *not by the process itself but by the relations which it bears to other processes* (Ashby 1960/1952: 6).

Thus, adaptation meant a change for the better in these relations. However, this still left the problem as to how an adaptive change could be evaluated as 'better'; for every evaluation was conditional and depended upon others.

Ashby wanted this evaluation to be done scientifically: he was a biologist who would not accept psychological concepts 'unless it can be shown in objective form in non-living systems; and when used it will be considered to refer solely to its objective form' (ibid.: 9). No teleological arguments were allowed. All nervous systems behaved in a determinate way; they obeyed the usual laws of matter. Such subjective elements as consciousness were to be left out because learning did not depend on them. That did not imply that consciousness was non-existent. Rather, it was even prior to all other forms of knowledge; but it could not be handled scientifically. 'Science deals, and can deal, only with what one man can *demonstrate* to another' (ibid.: 12). According to Ashby elementary physical-chemical events were adequate for the description of all significant biological events and could be treated as (numerical) variables. He could take this position because he restricted himself to behaviour. He needed feedbacks to describe the relation between organism and environment; some systems even appeared to be intelligent; they returned to their state of equilibrium in spite of obstacles. Moreover, they could react to variables with which they were not in direct contact. In order to be 'adaptive' a system had to change its internal organisation. Ashby constructed such a system, an 'homeostat'. It could adapt to external stimuli, and so demonstrated an elementary form of self-organisation.

4.3.4 The computer hardware metaphor and logical positivism

As we have seen, the pre-1956 frame was the outcome of an interaction process between several disciplines. In 1944 the hardware metaphor emerged for the first time as a shared idea as a result of a stabilisation process which appears to have been a kind of two layer process. The first layer was represented by the leading actors — where disciplinary traditions met.[19] We have seen that the education of most of them was not monodisciplinary. McCulloch, for example, was trained in logic and neurophysiology; von Neumann in logic and mathematics (Figure 4.2). The second layer consisted of the interactions during their work. Von Neumann was strongly influenced by the neurophysiological ideas of McCulloch, and by the construction of the digital computer EDVAC. All these and other interactions led to the shared opinion that there was a strong comparison between the brain and the hardware of a computing device. Both were logical devices.

However, the agreement did not go much further than that because of a double disagreement. On the one hand, there were people who were convinced that their approach could explain the *whole* of human behaviour. McCulloch expressed this view for digital devices, MacKay for analog ones (although on close inspection, they both made reservations). Middle positions were also taken — for example, Ashby did not want to go further than the simulation of adaptive behaviour. On the other hand, there was the split between the appropriateness of digital versus analog devices. Again, middle positions were taken: von Neumann clearly saw the need for the presence of both types for the explanation of human behaviour. Not surprisingly, the digital approach was adhered to mostly by neurophysiologists and logicians. For them the on-off, or 0–1 scheme was part of

Figure 4.2: Disciplinary interaction

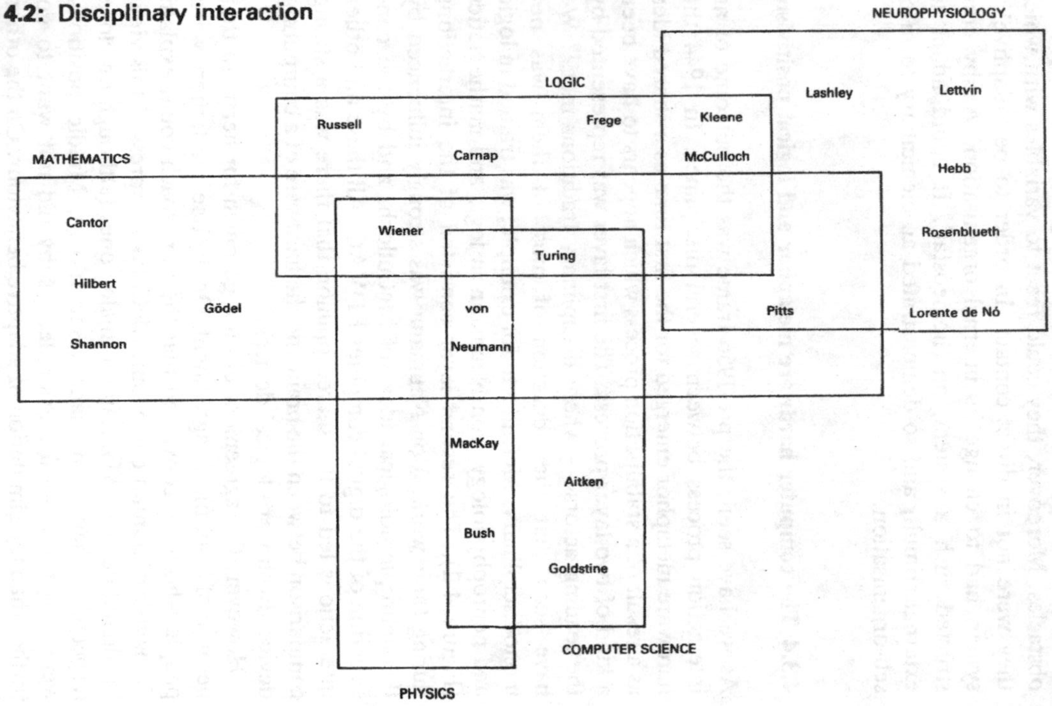

NEUROPHYSIOLOGY

Lettvin

Hebb

Rosenblueth

Lorente de Nó

Lashley

Kleene

McCulloch

Pitts

Frege

LOGIC

Carnap

Turing

Aiken

Goldstine

COMPUTER SCIENCE

Russell

von Neumann

Wiener

MacKay

Bush

PHYSICS

Cantor

Gödel

Hilbert

Shannon

MATHEMATICS

their tradition. However, physicists such as MacKay (who shifted later to a middle position) and the builders of technical devices had difficulties in thinking in on-off schemes; they were used to working with continuous quantities (notwithstanding quantum mechanics). As time went on, one can see that the position of logic was changing. Of course, logic must provide the answer, von Neumann stated, but he was also sure that none of the known types would do.

As I suggested earlier, what drew the different positions together was the computer hardware metaphor which had become digital, notwithstanding the changed position of logic correlated with the 1–0 scheme of the nervous system. Apparently, the technical success of digital computers was sufficient to outweigh scientific anxieties. The metaphor was also in agreement with the *Zeitgeist* which 'allowed' only logic and empirical studies and which was the dominant tradition, particularly in those Anglo Saxon countries where all these studies had occurred. That tradition — logical positivism — was certainly not limited to such disciplines as mathematics, logic and neurophysiology: the philosophical scene in those countries was also dominated by it. Thus, for example, the work of Carnap, which influenced so many of the AI people, began with the dogma that only contributions by logic and sense input could be allowed in a proper science. The same mentality could be found in most of the humanities in these countries. This tendency was strengthened by the strong desire for these disciplines to work as scientifically as physics; it was felt that they were impeded by the lack of an approach of their own which could similarly prove its validity. In psychology, for example, introspection had failed to provide any such solid base, promoting even a disgust for the use of a non-observable terminology. Only what could be seen — that is, behaviour — formed a legitimate object of study. This will be illustrated by the case of the Harvard psychologist, Boring.

4.3.5 Edwin Boring

Edwin Boring began his paper 'Mind and mechanism' (1946) by stating that he did not know what truth and reality were but that he did know what scientific truth was: the conclusion of inference from data. As such it had no certainty, but it had to

189

do for the moment (that is, it was pragmatically true). In psychology, the mental capacities of an organism were known when its properties were known. Operationalism, he said, was not a method but a 'frame of mind. It is a distrust of the use of words without knowing their referents' (Boring 1946: 175).

For Skinner, the properties of an organism were simply the ways in which stimulus and response were related. Thus, the way to think about psychological properties was thought of in terms of what could be expressed objectively in stimulus-response relations. From this perspective one could think about organisms only as either machines or robots, and in accordance with the spirit of the time, Boring used the latest technical developments: electronic brains.

It will surprise no one that he had no answer to Wiener's question about whether or not there was any human mental function which could be duplicated with electronic devices. In his paper he took up the question as to whether human capacities could be imitated mechanically. His answer was that they could — for example, by considering *attitude* as an operating selector mechanism (*learning* was not well understood). This strengthened his point that 'You could not understand a psychological phenomenon well until you can generalize the mechanistic properties which are implied by those operations by which the phenomenon is known' (ibid.: 185). Further, *symbolic processes* were also used by animals as well as humans — for example when communicating with an experimenter. 'That is why Pavlov called the gastric juice "psychic juice", because the animal spoke to him in the language of gastric secretion' (ibid.: 185). No intention was required for communication as long as it took place; *insight* had as its simplest form perception of identity; *introspection*, consciousness, could be defined as the ability to react discriminatively. 'I myself rather like to note that a falling stone is similarly aware of the gravitational field in this sense of reacting differentially to it' (ibid.: 189). If a robot could react to its own reactions, then 'you have introspection going on' (ibid.: 190).

According to Boring the robot gave the certainty 'to leave out the subjective, anthropomorphic hocus-pocus of mentalism. There is nothing wrong with mentalism if it uses rigorous definitions of terms, but usually it does not' (ibid.: 191); or as he stated it differently: 'Certainly a robot whom you could not distinguish from another student would be an extremely convincing demonstration of the mechanical nature of man and of the unity of the sciences' (ibid.: 191). This was a theme later taken up by Turing (1950).

Boring was quite aware that the functional resemblance between robot and human being still left the problem of the difference in underlying processes. For that reason, the robot was only of temporary assistance. It simply carried an argument against mentalism; the real answer was to be found in physiology. This approach to the study of human behaviour did not remain unchallenged. Against the dominant tradition based on the natural sciences (in a positivistic sense) an alternative was formulated, this time by people working in the humanities.

4.4 THE COGNITIVE REVOLUTION

4.4.1 Introduction

In some respects, one could speak of a single computer-metaphor frame. For many years there had been strong interactions between all the interested people concerned. Clubs had been set up to further these contacts: for example, the Teleological Society in the USA started in December 1944, and the Ratio Club in the UK in 1949. There had also been several conferences. In this context, the meetings of the Josiah Macy Foundation (since 1946) and the Hixon Conference in 1948 have already been mentioned. Building on this tradition, a conference was organised by Shannon and McCarthy for the summer of 1956 at Dartmouth College: it was entitled 'Automata Studies'. In general terms, the speakers and topics represented the pre-1956 tradition. MacKay's contribution has already been mentioned. It was part of the section 'Synthesis of Automata' which dealt with a simulation of the operation of a living organism. The other two sections dealt with Turing Machines, and finite automata, i.e. neural networks.

Notwithstanding all of these 'standard' contributions, the Dartmouth Conference was later seen as a turning point; for during the proceedings another approach was presented which must be seen as a break-through. It occurred in the form of a demonstration by Herbert Simon and Allen Newell (from the Rand Corporation) of a computer program which could prove some of the theorems in Russell and Whitehead's *Principia mathematica.* This program, 'The Logical Theory Machine', started what might be called a new era because it was based on a new approach. Statistically grounded calculating hardware devices no longer formed the point of comparison for the operation of the human mind but, rather, a program performing operations on symbols.

Not surprisingly, it was the younger members at the conference who pushed this new idea, and they came from different backgrounds. Simon and Newell, for example, were working for the Rand Corporation on human decision-making. Nevertheless, they too were educated in mathematics and logic. Those two components, psychological problems and logical tradition, set the scene for the AI developments to come. Coincidentally, on the same occasion McCarthy invented the term 'AI'; and though going against the grain for many people, it has nevertheless prevailed.

As already indicated, the period which started with the Dartmouth Conference is much better known. Therefore, less attention will be given to it here than to the pre-1956 era. After some introduction to this frame, I will focus on the systematic approach of Simon. Then Boden's standard text will be used to explain why so many students in the humanities are much more interested in AI than in any of the work of the preceding period. (In the whole of this section, one should keep in mind that AI is taken here as a univocal phenomenon. Actually, the AI frame is also the outcome of interactions of opinions but the details will not be spelled out here.)

Those activities which we describe as higher intellectual or even creative, such as problem-solving or common-sense reasoning, are immediately characterised by some sort of order. Thus, for example, when we have to solve a problem we do not start blindly, but apply some sort of heuristics:

> By 'heuristics' we mean to refer to things related to problem solving. In particular we tend to use the term in describing rules or principles which have not been shown to be universally correct but which often seem to help, even if they may also often fail (Minsky 1959: 36).

In general, one could say that the adaptive techniques developed so far in the pre-1956 frame were too clumsy, too rigid, at least to deal with this type of process. Better techniques were needed, more flexible ones which allowed faster adaptations.

In addition, amongst network-modellers there was an evident need for organisational hierarchies. Some evidence had appeared to justify 'cell assemblies' (Hebb), but

> once we have mechanisms for the formation of simple concepts as physical sub-structure, the remaining heuristic theory would not be very different from the kind concerned with the formal or linguistic models. One might soon need have little concern with the underlying nets (ibid.: 25–6).

An additional argument for working at a program level was the fact that programs themselves were many times easier to change than any technical lay-out of computing devices (or even neural nets). The difference of approach can be demonstrated for the case of pattern recognition. A matching of templates is less flexible than a comparison of lists of properties, and in particular when those properties are correlated.[20] A classical example of this is the program which analyses two-dimensional drawings of three-dimensional geometrical objects. Physical properties of the blocks constrain the huge number of *a priori* possibilities to an often unique solution.

4.4.2 Herbert Simon's systematic approach

Herbert Simon has developed a systematic approach towards AI. He states that both natural and human-made (artificial) objects have their own structure and their relation to the environment. Some of their internal structure can often be explained by the function or purpose they have with regard to the environment. One can also reason the other way around; but one has to keep in mind that objects with a different structure can have the same function. This gives us the opportunity to construct devices with the same functions as existing (natural) things. In this respect the computer is an ideal tool. 'Because of its abstract character and its symbol-manipulating generality, the digital computer has greatly extended the range of systems whose behaviour can be imitated' (Simon 1982/1969: 17).[21]

Of course, all comparisons are partial; they depend on which functions are compared. This means that it is easier to compare cases in which only the structural level, and not the actual details, have to be taken into account. What makes the computer so handy is the fact that it is a 'formal' machine. It operates quasi independently of its components.

> Here we are aided by the knowledge that *any* computer can be assembled out of a small array of simple, basic elements (. . .) The parts could as well be neurons as relays, as well relays as transistors (. . .) The possibility of building a mathematical theory of a system or of simulating that system does not depend on having an adequate microtheory of the natural laws that govern the system components (ibid.: 23).

All this sounds straightforward: it gives the impression that the problems have been brought to some higher level where they can be solved properly. However, even if all problems could be brought to that level, it would still be largely wishful thinking. There is no proper theoretical base for computer science or AI: in practice, it is to a great extent trial and error. Thus, for instance, in the case of the development and improvement of time-sharing systems, theory could perhaps have anticipated the experiments and made them unnecessary, but that did not happen.

Nearly all of this could have been said by neurophysiologists but there was now a new emphasis: 'The computer is a member of an important family of artifacts called symbol systems, or more explicitly, physical symbol systems' (ibid.: 26–7). Human mind and brain is another important member of that family. One has to be aware of the fact that these symbol systems are different from the ones mentioned in the pre-1956 frame as can be concluded from Simon's side remarks. They are mental representations which are used for all actions that are to be taken. They can be compared with such symbol systems as logic or language.

> Symbol systems are almost the quintessential artifacts, for adaptivity to an environment is their whole *raison d'être*. (. . .) Symbol structures can, and commonly do, serve as internal representations (e.g. 'mental images') of the environment to which the symbol system is seeking to adapt (. . .). Of course, for this capacity to be of any use to the symbol

system, it must have windows on the world, and hands, too. It must have means for acquiring information from the external environment that can be encoded into internal symbols, as well as means for producing symbols that initiate action upon the environment. Thus it must use symbols to *designate* objects and relations and actions in the world external to the system. Symbols may also designate processes that the symbol system can interpret and execute. Hence the programs that govern the behaviour of a symbol system can be stored, along with other symbol structures, in the system's own memory, and executed when activated (ibid.: 27–8).

It is this type of symbol which characterises humans.

4.4.3 Margaret Boden's popularisation of AI

Margaret Boden, a psychologist at Sussex University, has been very influential in popularising AI. Her book, *AI and natural man*, provides a survey of the field. She defines AI as the study of computer programs, or better: 'the use of computer programs and programming techniques to cast light on the principles of intelligence in general and human thought in particular' (Boden 1979: 5). The full emphasis here lies on the software side of the computer, and in particular on the ability to manipulate symbols. Symbols are seen here as purely conventional signs which, on the one hand, can be manipulated formally by a machine; and on the other, can be interpreted by a user. Due to the fact that the manipulation of symbols is completely formal, it is independent of the type of machine. The latter merely has to be a device which can handle symbols formally. Boden puts so much emphasis on this point because, in her opinion, AI gives psychology the opportunity to free itself from the devastating influence of the natural sciences. According to her, although psychology should not be reductionist in the behaviourist sense, it could be mechanised.

The orthodox positivist approach of psychology had assimilated too much of the natural sciences.

Whether a psychological phenomenon be categorized as 'inner thought' or as 'outer behavior', it must be conceptualized on the model of meaningful action on the part of a subjective agent rather than as a causal process in the natural world. Psychology must give an account of the meaning, or intentionality, intrinsic to mental life, and must regard the wider theoretical implications of any such account (ibid.: 395–6).

This emphasis on meaningful action by a subject does not mean that his/her mind cannot simultaneously be conceptualised as utterly dependent on 'a mechanistic causal system, the brain'. The gap can be bridged by AI because it deals with information at three independent levels: the machine level, which is determined by the properties of its hardware; the coding level, where the well-formedness of sequences of symbols is fixed; and finally, the semantic level, which is determined by the user of the information.

Boden goes along with Feigenbaum's definition: 'Intelligent action is an act or decision that is goal-oriented, arrived at by an understandable chain of symbolic analysis and reasoning steps, and is one in which knowledge of the world informs and guides the reasoning' (ibid.: 422). The use of computers forces the psychologists to make explicit the internal processes they suppose to be involved in human behaviour. Thus, these processes can be formulated algorithmically, giving the users the opportunity to test their ideas by running these programs and by comparing computer behaviour with that of human behaviour. Thus, their psychological theories consist of computer programs, while as a software processor the computer provides a suitable tool; psychologists handle the analogy by using software.

According to Boden, when one evaluates programs one should keep in mind that even in the case of ideal programs, intentionality will not be 'visible' on levels other than the psychological one. She maintains that although actual programs have some sort of intentionality, they fail at many points to match human performances, but this shortcoming might be only due to current programs. As long as 'the crucial notion in understanding how subjectivity can be grounded in objective causal mechanism is the concept of an internal model or representation, which we have seen to be central to many schools of psychology and AI' (ibid.: 428), one is on the right track. In Boden's perspective, previous trials were inherently insufficient. Although the cybernetic concept of negative

196

feedback (and its biological equivalent, homeostasis) provided a stronger analogy to genuine goal-directed action than did the concepts of seventeenth- or nineteenth-century mechanism, it was not sufficient to account for the intentionality of everyday behaviour. Boden has a higher esteem for AI, which, she believes, has a role to play in helping us to understand what it is to be a human being — though there are 'phenomena whose complexity makes it highly unlikely that adequate simulation will ever in fact he achieved' (ibid.: 444).

4.5 TRANSITION

The hardware metaphor was dropped in 1956 for its software version and this has since been seen as a real change of direction; in the eyes of many it constituted a revolution. As I argued at the beginning, although the date of birth of AI seems to have been given official sanction, it did not fall from heaven. Resistance to the prevailing methodology had been growing in several disciplines: for example, in linguistics, philosophy, psychology and computer science. For all of them 1956 happens to be the magic date. In addition, computer technology had made rapid progress: the realisation of stored programs made software manipulation a real option. In this section, attention will be paid to developments in some of the areas mentioned because they demonstrate how, at a disciplinary level, the new approach — AI — was anticipated. This will thereby establish the reason why it is unjustified to claim a revolutionary start for AI.

4.5.1 Rudolf Carnap and the retreat from logical positivism

In philosophy, logical positivism received a severe blow in 1956 when Rudolf Carnap published his paper on theoretical terms. It was his final step in an almost 30-year-long process of moving away from the positivistic dream he had written of in his *Die logische Aufbau der Welt* — namely, that logic and sense impressions should form a sufficient base upon which to build the whole of science. However, in taking science seriously, he had been forced to give up his *a priori* assumptions one by one. The last step occurred in the 1956 article, wherein Carnap was forced to conclude that theoretical concepts could have a meaning without any direct connection with sense impressions. They were to be given a separate status in their own right.

4.5.2 Karl Lashley and neurophysiology

At least as impressive was the 'rebellion' in the very heart of the neurophysiological camp itself. Karl Lashley (1898–1958) felt that their approach was unable to deal with large areas of human behaviour. In his contribution to the Hixon Conference, Lashley formulated his scepticism about the appropriateness of a bottom-up approach for the explanation of certain features of human behaviour. The work in neurophysiology, as done by McCulloch, was just such a bottom-up approach. The starting point was the neuron, and the complexity of the interactions (networks) of neurons was seen as determining the behaviour of an organism. Lashley also believed

that the phenomena of behavior and of mind are ultimately describable in the concepts of mathematical and physical sciences. In my discussion here, I have deliberately turned to the opposite extreme from the neuron and have chosen as a topic, one aspect of the most complex type of behavior that I know; the logical and orderly arrangement of thought and action. (. . .) My principal thesis today will be that the input is never into a quiescent or static system, but always into a system which is already actively excited and organized (Lashley 1951: 112).

Characteristically, he took language as his starting point. Each language had its own predetermined, orderly sequence of actions, but this and other phenomena (which had been neglected by neurophysiologists) could not be explained in terms of succession of stimuli. The only available theory explained coherence by postulating chains of reflexes/associations. However, this position was untenable because a great deal of evidence had accumulated to the effect that thought had a very complex structure: for example, there were reasons to believe that in the production of speech, at least three major neurological systems were involved. Even the movements for pronunciation had no intrinsic order of association. 'The order must therefore be imposed upon the motor elements by some organization other than direct associative connections between them' (ibid.: 115). The same problem arose for grammatical structure, whose order seemed to have its origin in the thought/idea expressed. However, the thought/idea itself did not have a temporal order, so there must have been an intermediate mechanism capable of transforming it into temporally ordered strings. Even typing errors indicated a complexity of 'order-forces' because parts of sentences were apparently anticipated. It was hard to explain these phenomena by appeal to the varying strengths of associative bonds.

Lashley paid so much attention to language because its organisation seemed to be characteristic of almost all other cerebral activities: these implied *serial ordering*. That neurophysiological research had thus far failed to provide a solution did not mean that it could not assist in finding one. 'Analysis of the nervous mechanisms underlying order in the more primitive acts may contribute ultimately to the solution even of the physiology of logic' (ibid.: 122). However, as matters stood, the behaviouristic explanation which rested on a simple stimulus-response scheme, was not merely inadequate, but was mistaken. In an organism's behaviour a stimulus seemed to play only a minor role. Furthermore, Lashley observed that the study of movement 'supports the view that sensory factors play a minor part in regulating the intensity and duration of nervous discharge; that a series of movements is not a chain of sensory-motor reactions' (ibid.: 122). In other words, movements of individual and groups of muscles are often anticipated, independent of any sensory control. Their co-ordination is not anatomically determined, but is a result of current physiological states.

Lashley pointed to the fact that the existence of temporal order requires some kind of scanning mechanism, but he had no idea what nature it might have. He attacked the prevailing idea that neurons were inactive most of the time and only active when they had to fulfil their specific goal.

The cortex must be regarded as a great network of reverberatory circuits, constantly active. A new stimulus, reaching such a system, does not excite an isolated reflex path but must produce widespread changes in the pattern of excitation throughout a whole system of already interacting neurons (ibid.: 131).

He had to conclude that 'the problems of the syntax of action are far removed from anything which we can study by direct physiological methods today, yet in attempting to formulate a physiology of the cerebral cortex we cannot ignore them' (ibid.: 134). Not taking them into account brought the risk of constructing a false picture. If there were processes which appeared to be inexplicable in terms of the contemporary construct of neurophysiology, then that construct was incomplete or mistaken. In his opinion, one had to start with a constantly active system, or rather, a composite of many such interacting systems.

Lashley found a willing ear at the conference. Apparently, time was ripe for his ideas, which had been the result of repeated failures to confirm simpler hypotheses.

> I agree with Dr. McCulloch that the transmission of excitation by the individual neuron is the basic principle of nervous organisation. However, the nervous activity underlying any bit of behavior must involve so many neurons that the action of any one cell can have little influence upon the whole. I have come to feel that we must conceive of nervous activity in terms of the interplay of impulses in a network of millions of active cells (ibid.: 145).

He ended his reply with the remark that real solutions had to be looked for: mere symbolisation would not do.

4.5.3 Noam Chomsky and the rejection of behaviourism

In what could be seen as a late echo of Lashley's plea, Noam Chomsky attacked behaviourism on exactly the same point: its inability to cope with the phenomenon of language. However, in a review of Skinner's book *Verbal behavior*, Chomsky went one step further: the behaviouristic approach was not only insufficient, it was inappropriate. Skinner had applied the behaviouristic approach to linguistic behaviour, his aim being 'to predict and control verbal behavior by observing and manipulating the physical environment of the speaker' (Chomsky 1959: 26). The only things which could be studied were 'observables' i.e. input-output relations, and functions which described the causation of behaviour. Even in speech, actual stimulation and history of enforcement were, as external factors, the overwhelming ones (as laboratory studies had revealed). On this view, the actual speaker had no real contribution to make.

According to Chomsky, these claims were far from justified. Results from laboratory studies could only be applied in a very gross and superficial manner to human behaviour. He considered in turn each of the central concepts which Skinner had used to explain behaviour and discussed their inadequacy. Drawing from linguistic examples, Chomsky showed that the terms by which Skinner wanted to show the lawfulness of behaviour — such as stimulus, response (strength), and reinforcement — were useless for the purpose of explaining verbal expressions. Thus, for instance:

When we read that a person plays what music he likes, says what he likes, thinks what he likes, reads what books he likes, etc., BECAUSE he finds it reinforcing to do so, or that we write books or inform others of facts BECAUSE we are reinforced by what we hope will be the ultimate behavior of reader or listener, we can only conclude that the term 'reinforcement' has a purely ritual function (ibid.: 38).

By applying these terms in this context, Skinner had deprived them of whatever scientific meaning they might have had under laboratory conditions: to fill them out with the full vagueness of ordinary vocabulary was of no conceivable interest.

In Chomsky's view, the behaviouristic approach was even inappropriate to explain *all animal* behaviour. He cited several studies which had shown that animals often acted in the absence of any reward. He also referred to 'imprinting', to the fact that behaviour could be an innate disposition to learn and to react later in life. This was most evident for Chomsky in the case of verbal behaviour.

Similarly, it seems quite beyond question that children acquire a good deal of their verbal and nonverbal behavior by casual observation and imitation of adults and other children. It is simply not true that children can learn language only through 'meticulous care' on the part of adults who shape their verbal repertoire through careful differential reinforcement. (. . .) It is also perfectly obvious that, at a

201

later stage, a child will be able to construct and understand utterances which are quite new, and are, at the same time, acceptable sentences in his language (ibid.: 42).

Here were processes at work which were quite independent of any 'feedback' of the environment. To talk, then, of 'stimulus generalisation', was only to perpetuate the mystery under a new name. Of course, processes such as reinforcement and casual observation had some importance, but any account of the development and causation of behaviour that failed to consider the structure of the organism could provide no understanding of the real processes involved. It was possible that linguistic features were not learned by reinforcement, but through genetically determined maturation.

As long as we are speculating, we may consider the possibility that the brain has evolved to the point where, given the input of observed Chinese sentences, it produces (by 'induction' of apparently fantastic complexity and suddenness) the 'rules' of Chinese grammar, (. . .) or that given an observed application of a term to certain instances it automatically predicts the extension to a class of complexly related instances. If clearly recognised as such, this speculation is neither unreasonable nor fantastic; nor, for that matter, is it beyond the bounds of possible study. There is of course no known neural structure capable of performing this task in the specific ways that observation of the resulting behavior might lead us to postulate; but for that matter, the structures capable of accounting for even the simplest kinds of learning have similarly defied detection (ibid.: 44).

For Chomsky it was evident that Skinner's application of behaviouristic terminology to verbal behaviour had failed completely. It was no more than a pseudoscientific apparatus, which was *less* suited to the task than the terms which he had wanted them to replace (such as refer, and intention, etc.). In Chomsky's opinion such efforts were doomed to fail at that moment in time; they were premature because language itself was not yet sufficiently understood.

For the time being, a precise preliminary account of verbal behaviour could not be given. What had to be done was to characterise it as completely as possible. According to Chomsky, it is reasonable to regard the grammar of a language L ideally as a mechanism that provides an enumeration of the sentences of L in something like the way in which a deductive theory gives an enumeration of a set of theorems. ('Grammar', in this sense of the word, includes phonology). Furthermore, the theory of language can be regarded as a study of the formal properties of such grammars, and, with a precise enough formulation, this general theory can provide a uniform method for determining, from the process of generation of a given sentence, a structural description which can give a good deal of insight into how this sentence is used and understood (ibid.: 55–6).

This meant that a sentence could be understood not because it matched 'some familiar item', but because it could be generated by a grammar. Only when this problem had been solved could the question be asked as to why such and such a sentence had been chosen. If one tried to do it the other way around, language would remain a mystery.

4.5.4 Jerome S. Bruner and psychology

Finally, attention will be paid to another discipline of the humanities, psychology. There also, attacks had been under way — for example, by Jerome S. Bruner at Harvard, against the prevailing behaviourism. The canonical position of behaviourism in psychology was characterised by four features; firstly, there was sensationalism: experience somehow copied the physical world; secondly, there was the empiricist credo — that learning about the world was only possible through experience which 'memorializes the coherence of stimuli' (Bruner 1984: 58); thirdly, objectivism, which characterised American psychology: data should be overt, public, non-subjective; and fourthly, physicalism: any psychological explanation should be physical and in the end biological. In short, good research had to be done in c.g.s. — that is, in centimetres, grammes and seconds.

When Bruner started his studies at Harvard in 1938, he found a psychology dominated by behaviourism, one of whose representatives was his teacher, Boring. Bruner noticed that 'the closer the study of perception got to the study of senses, the more acceptable it became to the behaviourist learning theorists' (ibid.: 67). For him the mystery of perception was not that our senses tell us so much about the world, but that we can do so much with the little information they provide. Evolution might have tuned our senses to the requirements of our habitat but they give us only a sample of it. Thus, he adopted his own position, and in 1946, together with Leo Postman, he started the 'New Look' movement which took errors (an embarrassment for psychophysicists) as the object of study. Errors in judgement of coins had pointed them in the direction of evaluating processes. 'The structure of appearance was shaped from the inside out, and not just from the outside in, from "sense data" into experience' (ibid.: 72). Perception was not so much concerned with things, as extensions, or colours, but with objects, events, and meanings. Statistical thresholds need not be studied, but rather, the question of what it means to 'see' something. To measure the time needed for this they used specially developed tachistoscopes. This led to the surprising result that unpleasant information was seen more slowly than pleasant information — a phenomenon which became known as the so-called perceptual defence. Another study, this time of values and interests, demonstrated 'semantic leakage', i.e. the influence of cognitive knowledge on perception. Although 'common' perception appeared more immediate than scientific, it was in fact the result of unconscious inferences.

> The irony is that it was not until we were able to look at perception as a genre of 'information processing' in the metaphor of a computer that we were able to see the *necessity* of its being the result of a prolonged process — for all its phenomenal immediacy (ibid.: 82).

Another dogma of behaviourism was the insistence that habits were determined by frequency. However, the New Lookers successfully attacked the frequency argument by experiments with unusually coloured cards. Some of their results proved to be more understandable when the theory of Shannon and Weaver became known. Moreover, in their struggle against the reigning behaviourism they also received support from Tolman at Berkeley and from W. Koehler who was visiting Harvard at the time.

In their turn, the New Lookers could not avoid the danger of becoming an in-group: 'self-centred, increasingly convinced that no one outside our magic circle had much to teach us' (ibid.: 93). However, in its reaction against radical behaviourism, the group did not succeed in formulating a theory of perception of its own: no one wanted to tackle that problem. Instead, they focused on social psychology and personality dynamics. That said, Bruner did continue with small-scale experiments. Starting from the notion that we do not see something unless we look for it, he formulated the idea that the world does not provide us with sensations but, rather, that it (mis)matches our hypotheses about what we are looking for by strengthening or weakening them. Hypotheses were generated by 'cognitive processes' which worked as models in the head. Sensations were changed into coding systems,

> the flow of information through them, the heuristics, and the clever ways in which that information was combined and recombined in the service of coding. All of it was extraordinarily primitive. But the rudiments of a new approach to perception were there (ibid.: 97).

Even within this new frame, Bruner retained his interest in that area of brain research which involved the study of perception by *topdown* processes. 'Tuning' became a central issue: an organism could prepare itself for events (sensations) to come. In the late 1950s Bruner left the area of perception for the study of thinking, and in particular inference and organisational strategies (which was until then, considered too mentalistic, too subjective, too shifty').[22]

Just at that time a new period in the study of perception had started which was largely a British contribution. They

> were considerably less interested in the philosophical problems of perception and more in their application to 'human performance' in practical tasks. They had to figure out what to do when machines (like fighter aircraft) outsprinted human reaction times, or when displays (such as early radar) spewed out information faster or more monotonously than most human operators could take it in and digest it. They called what they did, the study of *attention* — which disembarrassed them from the outset of some antique philosophical ghosts (ibid.: 101).

Bruner's contribution to the emergence of the new era was shaped into the form of a book he wrote with Goodnow and Austin: *Study of thinking* (1956). According to them the 'heart' of psychology was the 'cognitive process' which distinguished humans from other information processors. The New Look took on institutionaL form as the 'Center for Cognitive Studies' which was founded by Bruner and George Miller in 1960. Helped by the academic expansion in the post-Sputnik days, it was able to flourish. In retrospect, Bruner has looked back with a smile on the years of the New Look and those following it:

> it is hard not to be bemused by the accidents, the false starts, one's own pigheadedness, even the self-imposed blindness of in-groups. Our band of worthies, the 'New Lookers', started out to liberate psychology from the domination of sense-data theory, the notion that meaning is an overlay on a sensory core. It was part, I have no doubt, of a broader and deeper cultural movement to change the image of man from a passive receiver and responder to an active selector and constructor of experience. No doubt, too, it 'succeeded' — and the quotation marks are not there in false modesty. For in another sense it failed. New Look metaphor did not change theories of perception, however much it readied the ground for the change. The 'cognitive revolution' changed them, and particularly the respected metaphor of information-processing automata — computers (ibid.: 103).

The Center for Cognitive Studies has been quite succesful, in the sense that its influence spread to many parts of the world. The real question is: did the attitude of psychologists really change? For one, G. Miller is not convinced that it did.

> What seems to have happened is that many experimental psychologists who were studying human learning, perception, or thinking began to call themselves cognitive psychologists without changing in any obvious way what they had always been thinking and doing — as they suddenly discovered they had been speaking cognitive psychology all their lives. (. . .) In my opinion, however, the use of these mentalistic terms is still constrained by a positivistic philosophy of science, so that now we have in effect an oxymoron: non-mentalistic cognitive psychology (ibid.: 126).

However, was the change a complete one, even for Bruner himself? He realises that they had done only a partial job, for they were also encapsulated in the dominant frame.

We were so keen to track our subjects doing their informational thing that we designed our experiments accordingly, with features in instances whose values could be easily computed. (. . .) So hooked were we on our formal logic that we never fully appreciated what we actually found in that more lifelike microcosm of the adult and children. And so the strategies that we found were in the end appropriate only to the world in which we forced our subjects to operate. (. . .) We were willing to take whatever artificialization was necessary in order to show that the human processing of information was of the form of a strategy governed by a higher-order principle. It was our way of substituting something more directive and longer-term for the usual psychological idea of a 'response' as the basic unit. Edward Tolman and John von Neumann were very much in our minds, as was Egon Brunswik — men who wanted to highlight the sequential, strategic side of human performance. And so we became so caught up in our austere text that we failed to see some intriguing things right under our noses, like the narrative concepts in those more natural experiments. But we were not prepared to pursue the more general point that a strategy would depend on the context of a task. (. . .) The rest is ignored or passed over. We 'construct' rather than test. It was there before our noses, but we were looking for something else (ibid.: 127–8).

Bruner also formulated his reservations about the new style of thinking:

I think I am suspicious of 'formal' models of human behavior — theories couched exclusively in mathematical terms or in abstract 'flow diagrams'. I have always been sympathetic to the metaphors of computation and information processing, but resistant to getting trapped in their necessary measurement constraints. Perhaps I feel that such systems of measurement trap you on their flypaper while you are still wanting to fly. Their precision exacts a very high price in the abandonment of imagination — eventually, no doubt, well

207

worth paying. But not now! I did not, in consequence, get deeply drawn into the Harvard-MIT 'cognitive sciences' network — the remarkable group of people who were pursuing formal ideas about information processing and computing and decision making. George Miller and Oliver Selfridge kept me abreast enough, I thought, I regret now not getting more involved earlier. I turned to more concrete ways of formulating my ideas (ibid.: 99).

These insightful remarks set the scene for an evaluation of the so-called 'cognitive revolution'. Thereby, they transcend the objective of this chapter which analyses the transition from the hardware to the software computer metaphor. The point that Bruner stresses is that the 'cognitive revolution' — which AI claims to have brought about — was in actuality not radical enough because it was unable to free itself from the other half of the positivistic dogma: the prevalence of formal thinking. At the moment this further move has hardly been considered.[23]

4.6 AN EVALUATION

Following a description of the pre-1956 frame and a shorter one of the AI frame, some of the developments have been sketched which led to the change of frame. An evaluation will now be made of the character of this change and I will suggest that the actual development also seems to contain a moral for how human intelligent behaviour should be studied (by AI).

According to the most commonly accepted view, the work done in the pre-1956 frame was useful at the time, but became outdated as soon as a better approach had been formulated — one which was considered more suitable for the explanation of human behaviour. In contrast, what must be stressed here is that in reality the situation was rather more complicated. One preliminary remark worth making is that the so-called 'cognitive revolution' was not so much a revolution, as an important change of approach. It really has constituted a *significant change*: the point of comparison between computer and brain shifted from the hardware to software. To treat an organism as a processor of symbolic information was quite a different approach to that of the neural network model of pre-1956. Furthermore, the curse on the use of theoretical terms had been

Figure 4.3: Comparison between the different AI frames/paradigms

period	name	direction of approach	paradigmatic example	method	computer technology	representatives	disciplinary background
1930s 1956	hardware metaphor	bottom-up	vision	statistical calculations	digital analog	McCulloch Wiener/v. Neumann Turing/Kleene	neurophysiology mathematics logic
1956	AI	top-down	language	algorithms	digital serial	Simon/Newell Minsky McCarthy/Nilsson	humanities mathematics logic computer science
mid-1970s	Connectionist Model	bottom-up	vision	statistics	parallel	—	neurophysiology mathematics computer science
1980s	?	top-down	language	hermeneutics	—	Winograd Bruner Watson	humanities

broken; and, mentalistic terms were now considered essential for the exploration of human behaviour. In particular, research on linguistic behaviour and thought contributed to the breakthrough (Figure 4.3).

That said, several reasons can be given as to why these changes cannot be said to constitute a 'revolution'. Firstly, it was only a *partial renewal*. Both the pre-1956 and the AI frame shared the same programmatic commitment to logic. With this consideration in mind, one could say, that *continuity* was at least as important as *change*.[24] This becomes even more evident when one considers a closer look at the other aspect of (logical) positivism, that of the dominance of the so-called empirical method. With Carnap, it was agreed that empiricism had to weaken its grip: terms which dealt with non-directly observable or theoretical terms (mentalistic terms) had become legitimate to use in a scientific approach; but what kind of terms were they? Did they not remain encapsulated in the positivistic tradition? After many years Bruner confessed that their approach had not been so much dictated by its object of study, as by the *Zeitgeist*. The preference, or better, the pressure to obey logic, mathematics, and even computer science, had blinded the mind . . . and had blocked the access to phenomena. Moreover, could it not be said that in many cases the changes were little more than verbal? With Miller one can ask how far things had really changed? In fact, there are now some cautious signs of a growing awareness that it is necessary to get rid of the positivistic inheritance.[25]

Secondly, the transition from the hardware to the software metaphor was not so much a revolutionary jump forward, but more like the *next step in an ongoing process*, a consequence of an internal dynamic in the actual research programme. As will now have become clear from the description of the pre-1956 frame, the work on neural networks started within the same intellectual climate that gave birth to behaviourism. Research led to the conclusion that the behaviouristic *a priori* study of stimulus-response relations (which were seen as the only legitimate means of providing insight into the behaviour of organisms) was completely mistaken. An organism had a life of its own; its behaviour could not be understood from the stimuli it received, which might even be of only minor importance. Thus, the internal dynamics of neurophysiological research at least prepared the way for AI. This was also promoted by

Figure 4.4: The transition from the hardware to the software frame

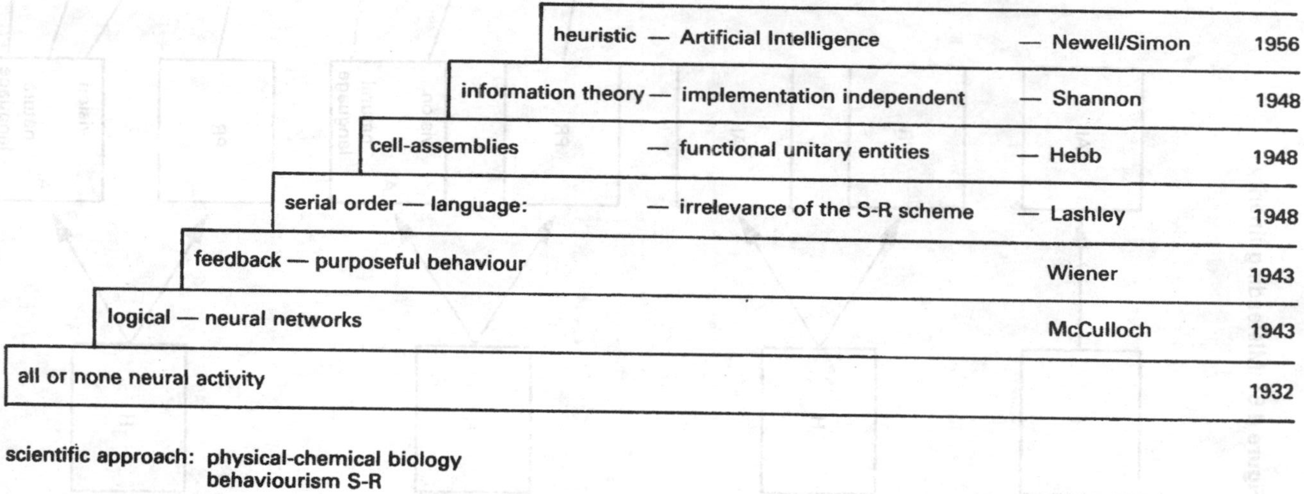

heuristic — Artificial Intelligence	— Newell/Simon	1956
information theory — implementation independent	— Shannon	1948
cell-assemblies — functional unitary entities	— Hebb	1948
serial order — language: irrelevance of the S-R scheme	— Lashley	1948
feedback — purposeful behaviour	Wiener	1943
logical — neural networks	McCulloch	1943
all or none neural activity		1932

scientific approach: physical-chemical biology
behaviourism S-R

Figure 4.5: Misleading linarity

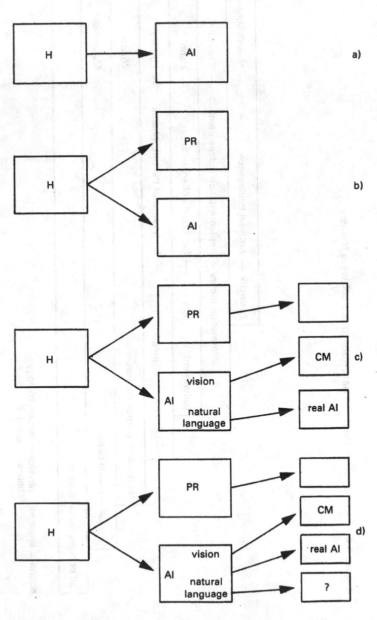

developments within such disciplines as linguistics, psychology, and philosophy (Figure 4.4).

Thirdly, it is difficult to speak of a revolution which outdated the previous approach (Figure 4.5a). Studying the claims of the AI frame one cannot avoid the impression that the step taken in 1956 was *not just a step forward.* In its process of stabilisation, AI has been unable to incorporate all of the valuable elements of its 'predecessor', some of which were mentioned at the beginning — for example, papers on neurophysiological work or parallel processing — but disappeared later on. This was not just a lagging behind due to the slow process of institutionalisation which must, initially, carry some 'old baggage' around. Changes take time, but the fact that this did not apply in every respect to the case here, becomes evident when one looks at the developments of AI during the 1970s and 1980s. After a period of euphoria in which the model of the human being as a symbolic information processor was paradigmatic for the explanation of every aspect of intelligent human behaviour,[26] one can see that some of the problems met in the realisation of the AI programme have pushed some people back to the once-despised network elements in the new Connectionist Model (CM).[27] Moreover, the hardware frame contained so many valuable elements that it has never disappeared. It even flourishes in the field of pattern recognition (PR) (Figure 4.5b). All this suggests that in the stabilisation process of AI, some elements have been eliminated prematurely — though it is an understandable course of events. As we have seen, new actors (even disciplines) succeeded in participating, while others were eliminated.[28] Thus, the outcome was bound to be different (Figure 4.6). It was not so much the case that other centres, e.g. the West Coast, took over; for in fact, it turns out that both frames have been strongly rooted on the East Coast of the USA, and in particular at MIT.

During most of this process, no agents outside the circle discussed here seem to have influenced the developments involved. As already indicated, there was the increasing need to do complex calculations which, initially, were demanded by electrical networks. For this purpose, Bush developed his analog computing device, Bell built a digital system for its telephone nets, and IBM was also interested. Later, however, during the latter half of the Second World War, all these initiatives became supported by the US Army which worked on

Figure 4.6: The research levels for SCOST

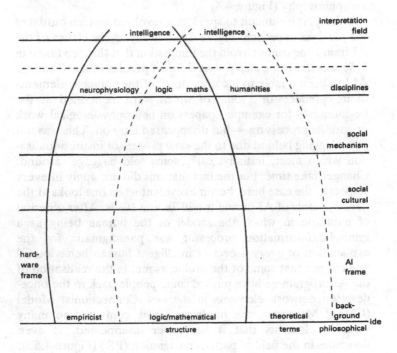

the construction of what has come to be considered as the first computer, the ENIAC. Before industrial goals and interests became apparent, the only fervent support in the USA came from the research department of the army, DARPA. Its long-term support for MIT became a tradition during this period. Amongst others, McCulloch's study on probabilistic logic was supported by the US Army, Airforce, and Navy as part of the work at the Research Laboratory of Electronics. The same applied to the work of Newell and Simon at the Rand Corporation. In addition, many other grants were given to special projects.

The Second World War had an influence which is not to be underestimated. Warfare push/pull in such technological developments as radar and the atomic bomb, but also the consequences of this impetus, gave rise to a change of scale which made it impossible for humans to cope with the many tasks. Amongst other things, the increased speed of aeroplanes made mechanisation of the anti-aircraft gunner necessary. Specific human actions had to be taken over by machines.[29] These developments posed scientific and technological problems of the mechanical simulation of tasks that were too fast for humans, as well as problems of human-machine interaction. This technological push/pull only speeded up after the War. It supplemented the increasing replacement of human tasks by machines. At the moment, the high expectations of AI within the SDI project scares many of its researchers.

From the publicity in our day, one might get the impression that the developments described here have given AI a solid place amongst the sciences, but that image is post-1981. Before this, only isolated individuals (and a few institutes) worked on AI, and influence remained restricted to academic circles. Of the participants mentioned in the first and second frame, all hold (or have held) chairs at America's most prestigious universities (with the exception of the UK participants). AI's frame has remained restricted to the circles of its participants. Outside this circle this type of work was probably considered as rather esoteric, if not to say futile.

Summarising, one can say that the cognitive revolution which forms the demarcation line between the hardware and software approach to understanding human behaviour, was mainly a methodological struggle about what was appropriate. The dominating scientific influence on the humanities has forced most of its American professionals to adhere to behaviouristic attitudes: only stimulus-response studies were carried out. Scientific studies of the neurophysiological base of behaviour have shown the deficiency of this scheme. An organism has its own role to play; it behaves actively. For the understanding of that form of behaviour, stimuli are only of secondary importance. This development, together with those in other disciplines, in moving away from behaviourism, has opened the way to a re-appreciation of the value of theoretical terms. Human behaviour should primarily be understood in mentalistic terms. However, this change in approach has not really been revolutionary: even if the primacy of sensation was broken, the other creed, positivism, was still around, with logic and mathematics as the scientific languages. Moreover, some lessons learned within the hardware frame appear to have been forgotten in the transition to the AI frame. At this juncture it looks as if they are going to have to be relearned. Not only has the hardware component been reintroduced, (particularly by the AI researchers working on robot vision) but the remaining positivistic attitude is under attack. The moral of the developments so far seems to be that a study of human intelligent behaviour has to take all aspects seriously — not only that part which can be described as the formal manipulation of symbolic information, but also those constraints which stem from its (hardware) implementation, as well as its meaning-giving part. Somehow, these aspects will have to be integrated, but their realisation in actual computer technology might prove them to be mutually exclusive. Thus, as matters now stand, the stabilisation process within AI is far from over. It will be interesting to follow its course.

NOTES

I would like to thank, in particular, the late Peter Boskma who gave me so much support in my research. I am also obliged to Barbara Saunders for the correction of my English.

All italicised words originate in the author's intention.

1. No effort will be made to cover the whole AI story; cf. P. McCorduck (1979), and more recently, H. Gardner (1985).
2. Cf. G. Boehme (1978) c.s., *Die gesellschaftliche Orientierung des wissenschaftlichen Fortschrittes* (Suhrkamp, Frankfurt); J. Schopman, 'Finalisation and functionalisation', *ZAWT*, vol. 11 (1980), pp. 347–53.
3. It thereby became more exact, in the sense of restricting it to group-characteristics (Harwood 1986: 182–3): but was it worth the cost? By reducing it to such group aspects the term lost most of its meaning.
4. Some of them are indicated in Figure 4.6.
5. Actually, the term 'technological frame' is used. The adjective 'technological' takes into account the influence of potential users on technological developments. For our purpose this adjective will be dropped because, as will become clear from our story, in the development of AI potential users do not play a comparable role. Until recently, one could not speak of a market pull. In the development of AI, future clients played hardly any role. As will be indicated later, the military played an important role, but the effects on AI were spin-offs of these developments: for example, the anti-aircraft gun as developed by Wiener and Bigelow; see section 4.3. Moreover, it is difficult to separate the technological and scientific components within AI's development.
6. This description lays emphasis on the field rather than on the

participants, but this is only done because of the need for clarity. The same story could also be told as the outcome of an interplay between actors.

7. Cf. McCulloch 1965/1943: 19.

8. An idea which is revived in Dretske, *Knowledge, or the flow of information* (MIT, Cambridge, Mass. 1981).

9. Cf. Jeffress 1951: 45.

10. Cf. McCulloch 1965: 152.

11. More recently, some doubts have been expressed about this claim. No independent proofs have been found that he indeed made these suggestions at the time (McCorduck 1979: 44).

12. Cf. Wiener 1962/1948: 269.

13. Cf. Jeffress 1951: 2.

14. This is one of the reasons why the digital devices became so successful. Their general-purpose character was another motive.

15. Cf. Jeffress 1951: 22.

16. Cf. Jeffress 1951: 23.

17. Cf. von Neumann 1958: 11.

18. Cf. MacKay 1956: 247.

19. A citation analysis clearly shows a reciprocal citation of the few people involved in the stabilisation process, with references by each of them to their own disciplinary background displaying little overlap.

20. For details of the example, see AI textbooks, e.g. P. Winston, *Artificial Intelligence* (Addison-Wesley, Reading, Mass., 1984).

21. Usually called 'simulation'.

22. Cf. Bruner 1984: 105.

23. One should mention here the work of, amongst others, Peter Wason, Johnson-Laird, and recently, Winograd and Flores. This trend is indicated in the last line of Figure 4.3.

24. This is even more so because behaviourism was not the only psychological approach around; cf. Boden 1979: 396.

25. Not only the authors mentioned in fn. 23 but many AI researchers also have their doubts about AI's claims.

26. Cf. Fleck's story about the dynamics of new theories (Fleck 1980: 42).

27. Cf. Special issue of *Cognitive Science*, vol. 9 (1985), pp. 1–169.

28. Schopman 1984: 92.

29. This worried Wiener to a great extent, as it still does scientists — cf. Weizenbaum (1986).

REFERENCES

Ashby, W. Ross (1960, orig. 1952) *Design for a brain: the origin of adaptive behavior*. Chapman, London

Bernstein, J. (1982) *Science observed*. Basic, New York

Bijker, W.E. (1987) The social construction of bakelite; towards a theory of invention. In W.E. Bijker, T.P. Hughes and T.J. Pinch (eds), *The social construction of technological systems*. MIT Press, Cambridge, Mass.

Boden, M. (1979) *Artificial Intelligence and natural man*. Harvester Press, Hassocks

Boring, E. (1946) Minds and mechanism. *Am. J. Psychology, LIX*, 173–92

Bruner, J. (1984) *In search of mind*. Harper & Row, New York

Carnap, R. (1964) The methodological character of theoretical concepts. In *Minnesota Studies in the Philosophy of Science I*, Minnesota University Press, Minnesota, pp. 38–9

Chomsky, N. (1959) A review of verbal behavior, by B. Skinner. *Language, 35*, 26–58

Feigenbaum, E.A. (1979) 'AI research: what is it? what has it achieved?. Quoted in Boden 1979: 422

—— and Feldman, J. (eds) (1981, orig. 1963) *Computers and thought*. Krieger, Malabar

Fleck, J. (1982) Development and establishment in Artificial Intelligence. In N. Elias, H. Martins and R. Whitley (eds), *Scientific establishment and hierarchies*, Sociology of the sciences, vol. VI, Reidel, Dordrecht, pp. 169–207; reprinted as Chapter 3 in this volume

—— (1984) Artificial Intelligence and industrial robots: an automatic end for utopian thought. In E. Mendelsohn and H. Nowotny (eds), *Nineteen eighty-four*, Reidel, Dordrecht, pp. 189–231

Fleck, L. (1980) *Die Entwicklung und Entstehung einer wissenschaftlichen Tatsache*. Suhrkamp, Frankfurt; English edition (1979) translated as *Genesis and devleopment of a scientific fact*. University of Chicago Press, Chicago

Gardner, H.(1985) *The mind's new science*. Basic, New York

Goldstine, H. (1973) *The computer from Pascal to von Neumann*. Princeton University Press, Princeton, NJ.

Harwood, J. (1986) Ludwik Fleck and the sociology of science. *Soc. Stud. Sci., 16*, 173–87

Heims, S. (1984) *John von Neumann and Norbert Wiener*. MIT Press, Cambridge, Mass.

Jeffress, L. (ed.) (1951) *Cerebral mechanisms in behavior: the Hixon Symposium*. Wiley, New York

McCarthy, J. (1959) Programs with common sense. In *Mechanisation of thought processes*, Her Majesty's Stationery Office, London, pp. 77–84

McCorduck, P. (1979) *Machines who think*. Freeman, San Francisco

McCulloch, W. (1951) Why the mind is in the head. In L. Jeffress (ed.), *Cerebral mechanisms in behaviour: the Hixon Symposium*, Wiley, New York, pp. 42–57

—— (1965, orig. 1943) A logical calculus of ideas immanent in nervous activity. Reprinted in *Embodiments of mind*, MIT Press, Cambridge, Mass., pp. 19–39

—— (1965, orig. 1948) Through the Den of the Metaphysics. In *Embodiments of mind*, MIT Press, Cambridge, Mass., pp. 142–56

MacKay, D. (1951) 'Mindlike behavior in artefacts'. *BJPS, 2*, 105–21

—— (1956) The epistemological problem for automata. In C. Shannon and J. McCarthy (eds) *Automata studies*, Princeton

University Press, Princeton, NJ, pp. 235–51

Kuhn, T.S. (1973, orig. 1962) *The structure of scientific revolutions*, Chicago University Press, Chicago

Lashley, K. (1951) The problem of serial order in behavior. In L. Jeffress (ed.), *Cerebral mechanisms in behavior: the Hixon Symposium*, Wiley, New York, pp. 112–36

—— (1958) 'Cerebral organisation and behavior'. In *The brain and human behavior*, Williams & Wilkins Comp., Baltimore, pp. 1–18

Mechanisation of thought processes (1959) Her Majesty's Stationery Office, London

Minsky, M. (1959) Some methods of Artificial Intelligence and heuristic programming. In *Mechanisation of thought processes*, Her Majesty's Stationery Office, London, pp. 3–28

Neumann, J. von (1951) The general and logical theory of automata. In L. Jeffress (ed.), *Cerebral mechanisms in behavior: the Hixon Symposium*, Wiley, New York, pp. 1–31

—— (1958) *The computer and the brain*. Yale University Press, New Haven

Newell, A. and Simon, H. (1972) *Human problem solving*. Prentice-Hall, Englewood Cliffs

Rosenblueth, A. Wiener, N. and Bigelow, J. (1943) Behavior, purpose and teleology. *Phil. of Science, 10*, 18–24

Schopman, J. (1984) The computational paradigm, a single one?. In *Proc. Sarton Conf.*, pp. 91–3

—— (1986) Artificial Intelligence and its paradigm. *ZAWT*, XVII, 346–52.

Shannon, C. and McCarthy, J. (eds) (1956) *Automata studies*. Princeton University Press, Princeton, NJ

Shurkin, J. (1985) *Engines of the mind*. Washington Square Press, New York

Simon, H. (1982, orig. 1969) *The sciences of the artificial*. MIT Press, Cambridge, Mass.

Turing, A. (1950) Computing machinery and intelligence. *Mind, LIX*; reprinted in A. Anderson (ed.) (1964), *Minds and machines*, Prentice-Hall, Englewood Cliffs

Weizenbaum, J. (1986) Ohne uns geht's nicht weiter. *Politische Monatschrift, 31*, Sonderdruck 332

Wiener, N. (1962, orig. 1948) *Cybernetics, or control and communication in the animal and the machine*, MIT Press, Cambridge, Mass.

—— (1966, orig. 1956) *I am a mathematician*. MIT Press, Cambridge, Mass.

Winograd, T. and Flores, F. (1986) *Understanding computers and cognition*. Ablex, Norwood

5

Involvement, Detachment and Programming: The Belief in PROLOG

Philip Leith

We have come to expect that studies in the sociology of science will show us the chameleon nature of scientific knowledge. As an early example, in the 1930s Ludwik Fleck dealt with the scientific fact, 'syphilis', and used it to contradict the then-current epistemological position that 'facts were facts'. After Fleck, facts were to be put back into the melting pot: they owed their significance as much to their history as to their current utility. The importance of this is that sociologists of science become socio-historians, forever linking present to past and, sometimes, past to present to future.

One mechanism for this long-term linking which Fleck suggested is the 'proto-idea', a sort of rough conception or belief from the past which turns up, in an amended state, in currently accepted facts. Fleck pointed out that the theories of the elements and of chemical composition, for example, were ideas which 'underwent a historical development from somewhat hazy proto-ideas'. It seems that this notion of proto-idea offers useful insights into the historical and social development of scientific ideas. What is not so clear is how it might be applied in the field of computing and Artificial Intelligence (AI) where proto-ideas can, as the example developed in this text will show, assume much more than a hazy correlation with currently accepted ideas. I shall present a proto-idea from the sixteenth century which has a striking counter in the artificial intelligence of the mid-1980s.

We can view proto-ideas within Elias's construct concerning the involvement and detachment of the individual and his or her beliefs (Elias, 1956). Although they are abstractions which were developed by different thinkers, there is a large area of

220

commonality between Fleck's proto-idea and Elias's position of involvement. Thus, the working over, the matching of proto-idea with empirical research in science, leads to the leaving behind of most of the involved aspects of the proto-idea (for example, the leaving behind of the notion of 'befouled blood'). This leaving behind of the less acceptable aspects of the proto-idea seems to be relatively permanent in esoteric circles of scientific development: we do not expect to see the ethical aspects of 'bad blood' return in future scientific research on syphilis, for such a return would be a step back to a more involved and less detached position. Such is the situation in the scientific field, but what of the situation in computer science and AI (where programming language design and function are two of the central concerns)? Are there any proto-ideas lurking in the background, in an area which is frequently perceived to be 'scientific'? I shall look at an example from the process of pioneering the use of a language (out of the many hundreds which are in existence): the claims made for the relatively new and successful PROLOG programming language. The claim being made for PROLOG is a relatively general one: it is one being made for computer languages which explicitly use logical methods as a means of representing the world.

In the example in this chapter, such a language has been used as a clarifying framework for the legal process and legislation. The proto-idea (of logic and law) can be traced back to Aristotle, though it reverberated substantially in the sixteenth and twentieth centuries, never quite having left the field at any point between Aristotle and the introduction of the computer. I shall argue that we cannot view either the Renaissance conception or the current conception of the logic/law thesis as being anything other than a highly involved position, and that, therefore, the belief in the utility of PROLOG must also be highly involved.

Why has the application of PROLOG (the logical computer language) to law been so forcefully put? Why have we not detached ourselves from the post-medieval proto-idea? One possibility, with respect to computing and AI, is that the design of programming languages is — by its very nature — dependent upon the involvement of the designer. I suggest that utilitarian factors (for example, speed of compilation, execution, etc.) are often of minor importance when a programming language offers a mirroring of a world view — thus, the development and acceptance of a programming language does not necessarily follow any foreseeable, unfolding path. The thought pattern of the group advocating any given programming language is often moulded in unexpected, and yet social, ways.

The chapter will deal first with Ramist logic — which, it will be argued, is the sixteenth century equivalent of PROLOG — and its application to the law; and then with PROLOG itself.

5.1 RAMUS AND HIS METHOD

Ramist logic is now little studied. It is seen by logicians (or rather, those who have heard of it at all — historians and philosophers of logic) as an aberration in the history and development of the discipline. Yet it is interesting to those researching the development of AI since it offers many commonalities with the development of this newer discipline. It is of just as much interest to those interested in the application of AI to law because one follower of Ramist logic, Abraham Fraunce, published a text in 1588 detailing the application of logic to law — *The lawyer's logicke*. One major problem, though, does exist for the investigator of Ramism: it existed at a time when Latin was the language of intellectual discourse. All of Ramus's texts (and those of his followers) were written in Latin, and even Fraunce's vernacular *The lawyer's logicke* contains long Latin passages. Researchers therefore have to rely upon the interpretation of others unless they have a facility with the Latin mode of medieval and post-medieval thought. For this chapter, then, Walter Ong (1958) and Perry Miller (1961) — who both present the fullest and most persuasive interpretation — are used.

We might consider Abraham Fraunce's thesis to be a proto-idea for some of the views which are current in AI: that, for example, we can produce formalisations of the law through logical methods which offer insights into what 'the law' actually is. We might do that; but we might also suggest that Fraunce's thesis is so close to current conceptions that it throws into doubt whether a view of logic and law from the sixteenth century can be so divorced from a view of 400 years later as to call it a proto-idea. We might suggest that the echoing of Fraunce's belief down the centuries has done little to change that belief: that today's belief itself is but that same proto-idea awaiting development, or discarding perhaps, in the future history of legal science.

The influence which Ramus had upon the post-medieval mind cannot be overstated; he was, to many of the time, sixteenth-century man. From the humble (though once noble) family of a French provincial charcoal burner, Ramus attended university — at the relatively late age of twelve — as the paid valet of a student with a more favourable background. Later barred from teaching or writing on philosophy, with his first two books being banned by the royal decree of Francis I, he was murdered on the third day of the Saint Bartholomew's Day Massacre in 1572 (some say by his rival). Ramus nonetheless managed to achieve a profound influence over a large proportion of his post-medieval colleagues: even those who were highly critical, it has been said, could not evade his influence. We might say that he influenced the thought style of more than his own supporters. From Aberdeen to Bologna to the New World, his influence was felt. Regarding the Ramist influence upon the New Englanders, we can quote Miller:

> The fundamental fact concerning the intellectual life of New Englanders is that they ranged themselves definitely under the banner of the Ramists. The Peripatetic system was indeed read at Harvard, but the Ramist was believed, and it exercised the decisive role in shaping New England thought. It is not too much to say that, while Augustine and Calvin have been widely recognised as the sources of Puritanism, upon New England Puritans the logic of Petrus Ramus exerted fully as great an influence as did either of the theologians (Miller 1961: 116).

Ong has pointed out that the vast majority of Ramus's followers were not great intellectuals — rather, that Ramism was congenial to 'impatient and not too profound thinkers', and he argues that Ramism offers important insight into current interests:

> Like a nerve ganglion, Ramism connects not only with readily discernible end-organs — explicit doctrines or theories of one sort or another — but also with more hidden, and at least equally important areas in Western culture,

alerting us to unsuspected connections between pedagogical developments and the rise of modern physics, between rhetoric and scientific method, or between dialectic and the invention of letterpress printing (Ong 1958: ix).

Ong wrote this before the widespread use of computers and before computational issues which went beyond the purely calculating into, for example, 'computational semantics'; and he also wrote this before the then-seedling conception of AI planted at the Dartmouth Conference of 1956 took root. It is my thesis that we can see Ramism evidenced, like an exposed nerve ending, in the collective thought of AI.

On the surface, Ramism is a form of logicism, yet it developed a logic which stands outside of any currently accepted logics. It has been described as a logical methodology with one of the most reckless means of quantification the intellectual world has ever known. It is best seen as a conceptual tool, a methodology, for describing the world — a dialectic which attempted to resolve the arts into axiomatic systems. Ramist logic was proffered as a way of coming to terms with the host of subjects taught in the post-medieval university. Ramus was not slow in showing how it could be used to formalise and encapsulate a myriad of subjects. His approach was to use his logical tool to axiomatise and reorganise subjects while he was actually learning them. Thus, for example, we have been told by his close friend and biographer, Nancel, that before he began work upon the Christian religion, he did not possess a Bible, only a New Testament. With mathematics he had problems, for he would often make mistakes when calculating with his abacus and get himself tied into knots. (A contemporary pamphleteer taunted him with a description of one of these moments when he had to 'crawl off the platform, mute as a fish'.) His confidence in his logical approach allowed Ramus to approach almost all areas. He even ventured outside of the arts subjects into theology, a venture which had often been met with accusations of heresy against those who tried it. Generally, Ramus felt that the power of his method overcame all borders, though he — as Nancel mentions — 'did little in law'.

That Ramus stood outside the Aristotelian framework of medieval scholasticism is his most important aspect, but just how colourfully he so stood is shown by Ong:

We are told that Ramus' public lectures were attended by huge crowds, at least at times, and his popularity with the teen-agers who made up his audience seems quite credible. Even the groundlings among the *bejauni* or 'yellow-beaks', as the newcomers at the university were called, would have their soul torn out of them by the histrionics of this man, whose remarks, whether concerned with a literary text or with dialectic or physics, are salted with constant denunciation of the authors his fellow professors were teaching and peppered everywhere with 'Holy Jupiter!' and 'Good God!' and who roared all this out· with the most spectacular declamation and gesture. Even in the days of the quarrelsome and excitable followers of Tartaret, Cranston, the Coronels, and John Major, logic had never been like this (Ong 1958: 34).

However, what of his work, the logic which made such a major impact upon post-medieval thought? This is perhaps less concrete than the man himself, changing with every edition of his works published, and using existing terms in confusing ways. Ramus stood in one corner of a triangular debate — that between theological scholasticism (represented by, for example, Thomas Aquinas), arts scholasticism and humanism. It is a debate which is much more complex than a paragraph or two can possibly clarify; but, very briefly and very crudely, we can give an indication of what had occurred. Theological scholasticism offered a very complex and theoretical form of logic. Contrary to the Kantian view that logic had not progressed one step since Aristotle, we can now state that medieval logic from theological scholastics had made substantial progress in a variety of directions: it had much in common with the formal mathematical logic of today. Henry (1972), for example, notes how many of the subtleties can only be drawn out by current non-classical systems (i.e. Lesniewski). As in the present, though, one of the problems of this medieval mathematical logic was that it had little relevance to the non-philosophical general thinker or scholar; it was too philosophical and its concerns were too incestuous - of concern only to the small circle who created and manipulated its systems.

Arts scholasticism was a scholasticism which differed vastly from that of the theological kind. Its concerns were much more practical and pedagogical, for arts scholars were teenage students at the medieval and post-medieval universities, the complexities of formal logical systems obviously over their heads. However, logic was still seen as one of the pivots of arts education. In order to make it understandable, it was taught almost by rote from textbooks such as Peter of Spain's *Summulae logicales*. Supposed to be a restatement of Aristotle designed for teaching, much of the *Organon* (the name given to the collected logical works of Aristotle) was omitted and various other elements added. Of most importance, though, was the idea given by this work that there was no difference between dialectic (i.e. argumentation using the tools of logic) and logic (i.e. the basis of sound reasoning). This lack of differentiation is important, because it was taken up by Ramus; thus, the term 'Ramist logic' might confuse those who see logic simply in terms of current mathematical logic or the more limited syllogism. The merging together of 'dialectic' and 'logic' is fundamental to Ramist logic. It was not to be the technical separation of argument and reason which Aristotle had made it in his work: rather, the two were to be brought together into one system where logic and expression were intimately linked into one. By this means, Ramus could transfer the certainty of a logical system to the confusions of the arena of argumentation. Thus, the problematical areas of the world could be seen in certain terms. As a system which was designed within the framework of teaching teenage university students, it allowed an admirable ease of expressing knowledge: everything was certain and simple. Thus, Ramist logic offered certainty and simplicity.

To Aristotle, dialectic was the proper use of logic; and, further, logic was analytica (Greek meaning 'to unravel'). Thus, logic unravelled and dialectics applied. The incorrect application of logic — seemingly discounted by the certainties of Ramist logic — was termed *sophistry* by Aristotle. To the certain and the simple, sophistry was anathema; hence, there was no mention of it arising from within the compass of Ramist logic. However, arts scholasticism was not simply a brutalisation of Aristotle, because it did borrow much from theological scholastic logic — particularly that of quantification, which appears heavily in the *Summulae logicales*; and it was a logic which concerned itself with epistemology and psychology, as do few of the mathematical logics of today. (Note how even Russell and Whitehead in their *Principia mathematica* tried to escape the problems engendered in providing an intuitive reading of logical 'implication'.)

It was in opposition to arts scholasticism that humanism — a school which incorporated both Ramus and Erasmus — and its version of logic grew. The main points which the humanists held against the logic of the scholastics were that it was arid, difficult and pedantic; Ramis's charge that 'ordinary people don't talk like that' perhaps sums up the criticism.[1] The aim of the humanist was to make logic practical and of use; take Ramus's own version of his turning from scholasticism:

> After my regular three and a half years of scholastic philosophy, mostly the *Organon* of Aristotle's logical works, terminating with the conferring of my master's degree, I began to consider how I should put the logical arts to use. But they had left me no better off in history, antiquity, rhetoric, or poetry. Thus I went back to my study of rhetoric, ended when I began my philosophy course four years before. My aim was to put the logical books of the *Organon* to the service of erudition (Ong 1958: 41).

And yet, since Peter of Spain's logical works were so influential in the teaching of students, the logic/dialectic which Ramus invented owed more to the rhetorical framework of the *Summulae logicales* than to the logical works of any of the Greek philosophers he cites as his sources. Rhetoric, we might suggest, is best seen as 'pleading' (incidentally, the Aristotelian *Treatise on rhetoric* uses the judicial analogy to introduce the notion). Ramism was most certainly a logic which was specifically rhetorical. Thus, we have a 'logic' which interrelates technical logic, dialectic and rhetoric. It is little wonder that some consider it to be a stepping away from the arrow-straight and unfolding path of 'logical truth'.

As to the method, Miller offers the best overview, and I will briefly provide a resumé of that here. The method made use of the diagrams which the new printing technology allowed the Renaissance writer to use; and, in fact, Ramus's method relies heavily on a diagrammatic technique, where the world is first divided into dichotomies and then resolved together in a dichotomous manner according to Figure 5.1:

> . . . showed that the material of any art, physics or medicine,

Figure 5.1

P. RAMI DIALECTICA.
TABVLA GENERALIS.

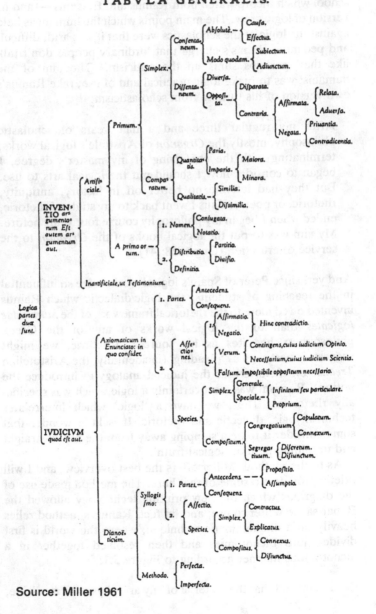

Source: Miller 1961

could be 'sown asunder' first by logic, first by a division of its two component parts, then by a subdivision of each part, and then by continued bifurcations of the subdivisions, until at last, on the right hand side of the page as shown in the diagram, the fundamental units, the indivisible 'arguments' would all be enumerated. Thereafter, by running our eye backwards along the brackets we can at once establish a proper classification. Taking our statement, 'fire causes heat', for instance, we find that this connection is absolute, agreeing, simple, and primary, whereas the statement, 'truth is opposite to falsity', presents a connection which is contradictory, negativating, contrary, opposite, and disagreeing, before it becomes, like the connection of cause and effect, simple and primary.

Ramus contended that this method was not merely simple and clear, but objectively true, that the content of every science falls of itself into dichotomies, that all disciplines can be diagrammed in a chart of successive foliations . . . (Miller 1961: 127).

Thus, the first part of the logical method was the step of invention, of discovering and classifying the arguments of the art or science. The second part was the judgement. Miller continues:

As with invention, judgement is both a transcript of reality and a process of the intellect. Things are 'glued' in nature, and the mind also glues one thing to another in thinking about them; if the mind is guided by dialectic it fastens things together in thought exactly as they are joined in fact. (. . .) The ways in which things are 'glued' can be classified into two main varieties, either in simple axioms, or else in discursive reasonings. We have two axioms, each composed of arguments, 'fire makes heat' and 'heat warms the body'. We put them together, exactly as we put arguments together to frame the axioms, and so achieve this discourse, 'When the body is cold it is wise to light a fire'. Hence Ramus dichotomized judgement into 'axiomaticum' and 'dianoeticum'. Each of these could then be subdivided; dianoeticum obviously consists of two kinds of discourse, for we may be able to carry our point merely by speaking one axiom after another, or we have to demonstrate it by the syllogism. We may be compelled to plead, 'fire makes heat, heat makes warmth, ergo, fire makes warmth', or we may, on so obvious a matter, fashion our oration by simply listing the axioms in intelligent sequence. Therefore Ramus dichotomized dianoeticum into 'syllogism' and 'method' . . .

But no matter how the divisions were conceived, procedure according to the Ramist logic was always the same: first we invent individual arguments, second, we dispose one with another to form an axiom; third, if in doubt, we dispose one axiom with another in a syllogism to get a conclusion; fourth, we set our conclusions in order and so make a discourse, a sermon, a poem, or an oration.

Of course, the procedure might be reversed: we may start with the finished sermon or poem, carve it into axioms, and resolve the axioms into their arguments (ibid.: 131-2).

Ramus was undeniably proud of the speed and simplicity of his method — he wished to have that fact noted on his gravestone.

5.1.1 Fraunce's application of the method

We shall not look in great detail at Fraunce's thesis because for current purposes it is only important as an example of the Ramist method, not in terms of offering us anything original. Ramus, of course, had considered law as a suitable area for his method (as his colleague and biographer, Nancel, stated), but had done little in that area. Fraunce, then, followed Ramus's intention.

Fraunce's work was moving into the post-Latin phase of scholarship: it was written in the vernacular (interspersed with legal French and Latin) and is thus slightly more easily accessible to today's sociologist. Much spleen is vented in the work against the Aristotelian arts scholastic, the remainder being an account of the usefulness of logic to the law:

> . . . I then perceived, the practise of Law to be the use of Logike, and the methode of Logike to lighten the Law. So that after application of Logike to Lawe, and examination of Lawe by Logike I made playne the precepts of the one by the practice of the other, and called my booke, The Lawyer's Logike . . . (Fraunce, 1588).

The main difference between Fraunce's use of logic in law and that of many of the contemporary proponents of logic and law (for example, see MacCormick, 1979) is that Fraunce was not too bothered about legal reasoning, i.e. how closely the reasoning of the judge or the lawyer is actually or should be 'logical'. Take the comments of Guest pointing to the current logical/law interest in logical legal cogitations:

What do we mean by logic? Too often discussion of this subject as centred around the rather barren controversy whether legal reasoning is deductive or inductive in form. . . . We must expect the position to be far more complicated. We must expect legal reasoning to be partly deductive and partly inductive, partly reasoning by analogy and partly the product of intuition, emotion or prejudice (Guest 1961: 181–2).

This is the position taken up by many researchers in the current climate of logic and law. Logic is thus an ideal form of reasoning and thinking to which the judiciary as arbiters of the legal process should work towards, freeing themselves as far as possible from any human failings (i.e. intuition, emotion or prejudice). We shall see that in the Kowalskian legal logic project this is not the case; but, rather, that this project is closer to Fraunce's.

Fraunce attempted to provide a framework for the transmission of legal information: he operated within a teaching framework and thus his logic was directed not towards the psychological aspects of judicial behaviour (that was more the interest of the arts scholastics), but towards the dissemination and classification of legal information. Even in the sixteenth century, the law was seen to be scattered throughout too many diverse publications; Edward VI was echoing a long-heard complaint when he stated, 'I would wish that (. . .) the superfluous and tedious statutes were brought into one sum together, and made more plain and short, to the intent that men might better understand them'. Fraunce, like every other improver of the law, points this out too:

But the law is in vast volumes, confusedly scattered and utterly indigested: so was all other learning not long ago: for this, though, do not blame the Law, I say, but lawyers

231

themselves who allowed the Law to be out of order because they never knew Method: No, blame neither Law nor ancient lawyers who due to the general misery of their obscure age, could not see everything, but exclaim against yourselves (i.e. the arts scholastics), who in this flourishing time and with blessed opportunity will see nothing. If those ancient fathers of our Law lived now things might be better, but if you had existed then things would have been worse. For neither can you do what you should, nor will you let others do what they would for the more orderly explication of the law (Fraunce, 1588). (my rephrasing)

The Ramist method, then, was to allow the law to be formalised into a more orderly, explicated body of knowledge. There is an element of 'methodological' arrogance (we might term it) in Fraunce's claim — he states that the problem of legal obscurity does not lie with the legal forefathers *per se*, but the problem lies with the fact that they did not have the Ramist method. With method, he claims, the ancient scholars could have achieved so much more!

To recapitulate, then, Fraunce was a university man (turned barrister), and was thus intimately interested in the pedagogical aspects of knowledge — legal information as we might now term it: how law is to be expressed, transmitted, and so on. In the text he does not really bother with how judges should 'think' or 'reason', so long as they use the method as a means of clarifying the underlying legal epistemology, that framework which exists as a part of the world and which can only be expressed through the method.

And we can see how this work of Fraunce — like much of the work carried out in the field of AI today — was a work of the young:

> Professor Howell has given an excellent account of the further spread of Ramism at Oxford and Cambridge in the late sixteenth and early seventeenth centuries. With the help of this account (. . .) it can be seen how juvenile was the pattern of Ramism in the English-speaking world. Although there is a great deal of interest in Ramism in England, with very few exceptions almost no one connected permanently with either Oxford or Cambridge writes on Ramism for his colleagues of the university milieu. There is considerable

reading of Ramist works by students, with or without direction, and a good deal of shouting for or against Ramism by sophisters or other youthful university disputants — indeed, there are recurrent references to the youthfulness of British Ramists (e.g.) Abraham Fraunce. (. . .) Oxford's most active Ramist, Charles (. . .) Butler, was famed not as a seriously scientific logician or philosopher but as an author of preparatory-school textbooks in rhetoric (Ong, 1958: 303).

Thus, there are some points with which we can briefly conclude regarding Fraunce's work: firstly, it offered little logical novelties, but did offer a novel application of 'the method'. Secondly, it was more concerned with passing legal information in a consistent format than with legal reasoning as a psychological issue. We shall see below how closely these points accord with those we might extract from an examination of an AI project of the 1980s.

5.2 KOWALSKI AND LOGIC PROGRAMMING

James Fleck gives us the origins of much of the current interest in the PROLOG programming language: 'R. Kowalski made a name for himself with his vigorous promotion of the predicate calculus as a programming language in its own right, an approach which became an independent strand of research termed "Logic Programming"' (Fleck, J., this volume: 128). We can now view this as, perhaps, an understatement simply by looking to the number of cited research papers which concern themselves with logic programming (some 1600 according to Balbin and Lecot (1985) — a huge number given the youth of the discipline) and the value of research monies being spent upon it both by academic funding agencies and commercial bodies. In many ways, logic programming — evidenced through PROLOG — is seen as the language of the future, and is being 'vigorously' sold as such. Take, for example, Kowalski's foreword to a *vademecum* for proto-logic programmers:

This book is a major contribution to logic programming. It sets out for the first time in one place a comprehensive yet accessible introduction to all aspects of our subject. . . . This beautifully written book will be a joy to both novices and experts. It will help waken the novice to the wider world of logic programming, and it will help stir the logic programming expert to greater understanding and further enthusiasm for our subject (Kowalski, in Hogger 1984: foreword).

We can also note one logic programmer's relating of how PROLOG reached Japan, indicating just how unexpected can be the turns of the computing world:

> Considering that PROLOG was relatively unknown to the computing world prior to the announcement of the Fifth Generation, the choice of that language as the basis for a national project of the scale proposed is quite a bombshell. Certainly, it seems to have come as a complete surprise to the logic programming community outside Japan, which had little idea of the Japanese interest in PROLOG and logic programming. (. . .) Apparently Fuchi had been interested in logic programming since reading Kowalski's 1974 paper. (. . .) There evidently must have been quite a lot of politicking to get Fuchi's 'bold proposal' accepted, but eventually, according to Furukawa, people were persuaded that in the 1990s 'even cats and spoons will write PROLOG' (. . .) (Warren 1981: 150).

It would obviously be interesting to see a fuller analysis of the historical development of logic programming, from its implementation in France by Colmerauer, through its acceptance in Japan and its later influence upon researchers in the USA where it has met opposition from LISP (a programming language which, like PROLOG, owes its theoretical foundations to that cross between mathematics and logic). However, such a study would miss out on seeing PROLOG and logic programming within the longer-term context: it would omit the fact that there exists a commonality with the sixteenth-century method of Ramus. We shall be concerned with the longer-term context. In particular, three interrelated points are argued. Firstly, that logic programming can be seen as an instance of 'method', with close analogies with Ramist 'method'. Secondly, the involvement which can be seen from proponents of logic programming mirrors the involvement with Ramist method. Thus, there is not only a close technical analogy between the methods of Ramus and Kowalski (i.e. as a means of encapsulating knowledge), but there exists a similar involvement from the proponents of both the positions. Thirdly, and following directly from the second point, we can conclude that the detachment which we would expect (given claims made for the nature of 'computer *science*') from a 'scientific' investigation of logic programming is missing. From these three points, we can therefore suggest that logic programming is, at best, surrounded by a sort of pseudo-detachment, and that programming languages are not immune from the pressures and problems of socio-cognitive life. However, it being important to address basic aspects, it is essential to examine briefly the nature of the PROLOG language itself before we try to draw analogies.

One of the marked points which is found by the observer who examines the research topics of those in the general field of computer science or the more specific field of AI is that a heavy preponderance of topics involve the application of mathematical techniques. We can, for example, point to projects attempting to 'prove' programs correct, projects attempting to produce 'metrics' to analyse 'good' programs, and projects attempting to specify 'programs' in formal and mathematical ways. The origins of logic programming lie in another application of mathematics to computing — that of deductive theorem proving or, less fashionably, deductive inference. One early commentator on the field of AI has described the latter as:

> Deductive inference, using the special symbols and procedures of mathematical logic, is a derivational approach that can be used in a wide range of problem-solving situations. Here a problem situation is represented by formal expressions that stands for some premises, and the particular problem is presented by a *theorem* to be proved from those premises. The proof method may be *semantic* relying upon the possible meanings of the expressions, or *syntactic*, consisting of abstract symbol-manipulation rules. In either case, it relies upon some established logical system and its associated *rules of inference* (Raphael 1976: 137).

Kowalski's PhD was in the area of theorem proving (Kowalski 1970), and PROLOG can be seen as an applications model of the more theoretical system. PROLOG consists of a series of causal rules, where those rules which only consist of conclusions are facts, and rules of inference which are based upon resolution theory. Thus, the programmer will simply list a collection of rules or facts; PROLOG supplies the inference mechanism; and the user of the program asks the executing program a series of questions which are to be proven. Kowalski himself gives the following example of its use in data bases:

> A program (or database) is a collection of sentences (also called Horn clauses) which express the information which can be used to solve problems (or to answer queries). A Horn clause sentence is one of the following:
>
> 1 A simple assertion, e.g.:
>
> John likes Mary
>
> Bob likes Mary
>
> Mary wants bicycle
>
> wheel is part of . . .
>
> 2 An implication, e.g. such statements as:
>
> Mary likes x if x likes Mary
>
> . . .
> . . .

Given the database/program in the two points listed above, the query:

Find x where x likes Mary

has two answers:

Answer: x = John

x = Bob   (Kowalski 1981: 79).

To PROLOG, the world is thus full of facts and assertions and a means of asking questions. PROLOG is not, incidentally, seen as the last word in logic programming — many are the attempts to improve it and extend it — but it is seen as one first example of a logic programming language. However, programmers in commerce and academia are not usually interested in who likes Mary; their concerns are much more practical. Thus, why should they expend the not inconsiderable effort in learning a new language?; and more so, why should they do so when it is commonly agreed (sometimes reluctantly, though) that *any* programming language can be made to simulate *exactly* the behaviour of any other programming language? Given this fact, why should PROLOG be perceived as offering anything which other languages cannot offer?

The fact is that computer scientists will argue that substantial progress has been made in the development of programming languages, even though both theoretically and practically, it can be argued that every progressive development can be matched by pre-progressive languages. The '*Lighthill Report*' (Lighthill, 1973) shows a case in point. Lighthill's suggestion that little had been achieved by AI researchers was met by responses (in that report) from Sutherland and Michie that progress had been achieved in the development of programming languages such as PLANNER (a language which was much designed but, to my knowledge, never implemented) and CONNIVER. The implication contained in their replies to Lighthill was that it was progress because the languages were moving away from the machine — thus they encapsulated knowledge in a more 'intelligent' manner. Michie was not slow in presenting economic arguments:

> The subject [that is, AI], in so far as it comes within the Computing Science Committee's realm of interest, is concerned with machines, and in particular computers, displaying characteristics which would be identified in a human being as intelligent behaviour. Perhaps the characteristics which are most important are those of learning and problem solving. The applied benefits which may be gained from work in this field could bring considerable economic benefit to the country. They are two-fold:
> a To relieve the burden at present on the systems analyst and programmer in implementing applications:
> b To enable new and more complex applications to be undertaken in this country in competition with work elsewhere (Michie 1973: 41).

Thus 'ultra-high level programming languages' (Michie's term) are to relieve the burden in (a). However, there are of course a variety of forms which an ultra-high level language can take —

237

progress might not necessarily be the unfolding of a set technological path. It could be argued that the major concern of the discipline of computer science is the decision as to just which path should be taken.[2]

The reader of Kowalski will see that logic programming offers an ultra-high level language which is particularly suitable for users:

> Employed as a language for communicating with computers, logic is higher-level and more human-orientated than other formalisms specifically developed for computers. In contrast with conventional computing methodology, which employs different formalisms for expressing programs, specifications, databases, queries and integrity constraints, logic provides a single uniform language for all of these tasks. (. . .) The meaning of programs expressed in conventional languages is defined in terms of the behaviour they invoke within the computer. The meaning of programs expressed in logic, on the other hand, can be defined in machine-independent, human-orientated terms. As a consequence, logic programs are easier to construct, easier to understand, easier to improve, and easier to adapt to other purposes (Kowalski 1979: preface).

Thus, PROLOG offers nothing computationally new (we might say) but does offer something 'interactively' new — it conforms most closely with the most suitable method for humans to programme computers. Progress, in the PROLOG context, has been seen as bringing computer programming towards the programmer and further from the machine; and logic, it is perceived, is closer to the programmer than any other representational method. To Kowalski, it is implicit that the world is a logical world, and that it is this aspect which allows PROLOG its supremacy as a programming language. Take the comments he makes on the use of logic programming in education:

> Logic is the single academic discipline which is common to all subjects taught in school. It provides a single uniform computer-intelligible language which is useful for expressing databases and programs for such diverse subjects as language, mathematics, the social sciences and the natural sciences.
>
> In mathematics, for example, transitivity of relations is just a

special case of recursion: x<y if x<z and z<y. Moreover, the problem of solving several equations in several unknowns, e.g. x+y=2 *and* x-y=0 can usefully be introduced to children as the problem of answering a database query.

In English, logic clarifies, for example, the meaning of the commas in the sentence

All students, who study logic, like logic.

With commas the sentences means

x likes logic *if* x is a student.
x studies logic *if* x is a student.

Without commas it means

x likes logic *if* x is a student and x studies logic.

The examples could be multiplied by considering all the other subjects taught in school. The development of logic as a computer language makes it especially appropriate, therefore, as a means of introducing children to computers. It contributes not only to the teaching of programming, databases and program specification but also to the application of logic to subjects traditionally taught in school (Kowalski 1982: 16).

It is a harking back to the cry of Ramus where: 'Dialectic is the art of arts and the science of sciences, possessing the way to the principles of all curriculum subjects. For dialectic alone disputes with probabilities concerning the principles of all other arts, and thus dialectic must be the first science to be acquired'. It is also a harking back, as we see below, to the concept of logic as method.

Just as the influence of Ramus cannot be underestimated (though it has been forgotten by many), Kowalski's additional influence on the academic and commercial world cannot be underestimated. From the publishing of his formative paper (Kowalski, 1974) to his *tour de force* of 1979 (Kowalski, 1979), the interest in PROLOG and logic programming has been impressive — some might say astonishing. It is an idea which seems to have found its time. There is no need here to list the influence which Kowalski's ideas have had: there are constant international conferences on logic programming, journals whose entire contents revolve around technical logic programming issues and much money is being spent by both the UK and Japanese governments on logic programming research. Whereas Ramus's influence lasted for quite a considerable period of time across great barriers of space (taking sixteenth and seventeenth century geography into consideration), Kowalski's influence and ideas have international effect — if not over a long time period, at least over large research funds (for example the dissemination of funding from Alvey and ICOT).

It would be possible to analyse, in terms of esoteric and exoteric groups, the make-up of the logic programming arena. One such esoteric group to be considered would have to be that group at Imperial College, London, where Kowalski is Professor of Computational Logic. Within the academic software teaching and research group, it seems — to the outsider — that the entire department is involved in logic programming. Some are involved in the theoretical issues (Clark and Hogger, for example) and some are involved in more practical issues (Ennals, Sergot and Hammond). To some extent, Kowalski appears to stand above the details of logic programming, leaving the particulars to the group. His role is that of advocate for logic programming, a role which he plays out through academic and commercial contacts and consultancies and through involvement in the provision of research funds as a member of an Alvey advisory committee. It would seem to be difficult for any member of that group to move away from such a logic programming hegemony, for a scientific establishment based upon that logic programming technique must be expected to control its members.

One aspect of Ramus and his logic was the belief that it was applicable across a wide area of academic disciplines — that it was the coherent structure which allowed analysis and development, if not the genesis, of knowledge. This aspect has its corollary in Kowalskian logic programming. We have seen, above, how he writes of logic as providing 'a single uniform language' for all the tasks of computing, and how it 'is the single academic discipline which is common to all subjects taught in school'. We might see such claims as the rhetoric of someone who is advocating a technique and wants to draw attention to it. However, this would be to underestimate the efforts which both Ramists and logic programmers actually put into proving these claims: they are not rhetorical statements; they are statements which can be substantiated.

The claims have been made for logic programming in education, for logic programming in theoretical computing, for logic programming in general computing, and — noted in this chapter — for logic programming in law. However, before we look at logic programming in the law, it is useful to note Turkle's argument that people in AI are intellectual imperialists (Turkle, 1984; see also Bloomfield, this volume, p.76).This description, it seems to me, might be applied to Ramus and his logic: note, for example, how he invented his method and then went out to use it on other disciplines. Kowalski, along with members of his esoteric circle, has also done this: take the case of logic programming and the law. Jurisprudence, the philosophy of law, has a history stretching back at least to John Austin who became the first (unpaid) professor of the new discipline in the early nineteenth century; and, of course, there have been many writers on the law for many centuries before that in both the Roman and European traditions. Yet the reader of research papers on logic programming and the law will notice that almost without exception the writers have felt confident that these philosophers of law can safely be set aside.[3]

Take the most prestigious article by the logic programming proponents on this subject (Sergot *et al.*, 1986) in the *Communications of the ACM* which contains 42 references: *none* of these refer to traditional legal/jurisprudential literature. Moreover, while it might be thought that research into systems to handle 'law' would be of concern to the British ESRC (responsible for the social sciences), this project was funded by the SERC (responsible for the hard sciences). Incidentally, Alvey also seems to have made funding decisions on legally orientated research without academic legal advice: the IKBS DHSS (British Department of Health and Social Security) demonstrator, for example, seems to have been conceived between logic programmers and administrators. We have, then, research into an area outside the usual ken of the researchers. (None of the Imperial College team on the logic programming project are legally trained or experienced — though one, at least, has considerable experience of administration of DHSS rules and regulations.)

In the next section we will look more closely at the legal side of the logic programmers research.

5.2.1 Logic programming and law

Work into the role of logic programming in the law seems to have begun with Hammond's representations of DHSS rules and legislation (Hammond, 1983), a project which was aided by an expert in these rules and legislative writings from that Department. From that beginning, a more ambitious project — the formalisation of the British Nationality Act (1981) — was begun; and it is this project which has achieved the most publicity with regard to that group's legal work. We shall therefore look at its goals and aims here, represented by its major paper (Sergot *et al.*, 1986).

The thrust of the British Nationality Act (BNA) project is to formalise the legislation relating to just who is a British subject and who is not. At its inception the Act was the subject of fierce debate, partly because it introduced barriers to 'full' nationality to those who considered themselves to so be: for example, those resident in Hong Kong, opponents to the Act claimed, became 'second class' British subjects. In the same vein, after the Falklands/Malvinas episode the Falkland islanders' position was amended so that they were to be 'proper' British citizens. There was also a debate about the confusion it would cause to those already in the country — whether they could remain here, whether children of immigrants would be able to claim citizenship, etc. It was also perceived as being a highly racist piece of legislation. It was a piece of law, then, which was in the public eye because of its contentious nature.

The team have written on why they chose this area as their test bed:

> The British Nationality Act 1981 was chosen for this experiment for a number of reasons. At the time it was first proposed, the act was a controversial piece of legislation that introduced several new classes of British citizenship. We hoped that formalization of the various definitions might illuminate some of the issues causing the controversy. More importantly, the British Nationality Act is relatively self-contained, and free, for the most part, of many complicating factors that make the problem of simulating legal reasoning so much more difficult. Furthermore, at the time of our original implementation (summer 1983) the act was free of the complicating influence of case law (ibid.: 370).

The formalisation of the Act was carried out by analysing the legislation and forming rules which could be fed into a PROLOG program. Obviously some problems were met — vagueness and imprecision being two of the most immediate:

> A complication that we anticipated was the presence of vagueness. The act contains such vague phrases as 'being of good character', 'having reasonable excuse' (. . .) The treatment of vagueness and case law is the subject of current investigation in our group.
> In addition to vagueness, legislation is generally thought to contain both imprecision and ambiguity. . . . In fact we found fewer such examples than we originally expected. In practice where imprecision or ambiguity did exist, it was usually possible to identify the intended interpretation with little difficulty (ibid.: 371).

As an example of the BNA formalisation, we have the authors themselves describing it:

> As an example of formalization using Horn clauses, consider the first clause of the British Nationality Act:
> 1(1) A person born in the United Kingdom after commencement shall be a British citizen if at the time of birth his father or mother is
> (a) a British citizen; or
> (b) settled in the United Kingdom.
> The act states that 'after commencement' means after or on the date on which the act comes into force.
> As a first approximation, 1(1)(a) can be represented by the rule
>> x is a British citizen
>>> if x was born in the U.K.
>>> and x was born on date y
>>> and y is after or on commencement
>>> and z is a parent of x
>>> and z is a British citizen on date y
> Here x, y and z are *variables*, which can have any values. For example, the common knowledge that a parent is either a father or a mother can itself be expressed by two rules:
>> z is a parent of x if z is mother of x
>> z is a parent of x if z is father of x　　　　(ibid.: 372).

243

THE BELIEF IN PROLOG

By a process of trial and error the rules are extracted from the legislation and inserted into the PROLOG horn clause format. The trial and error is required because later parts of the legislation are often needed fully to formalise earlier parts. Also, PROLOG itself is not the ideal legal logic programming language:

> We will see later that this formalization of 1(1) is inadequate; partly because of the need to determine the date on which an individual acquires British citizenship, and partly because elsewhere in the act it is necessary to know the section by which an individual is deemed to be a British citizen. We will also see that Horn clause logic itself is not entirely adequate for representing legislation in a natural manner and that, in many cases, Horn clause logic must be extended to allow negated conditions in rules. The resulting fragment of predicate logic will be called *extended Horn clause logic* (ibid.: 372–3).

As to the usefulness of such a formalisation, there are two main uses. Firstly, it can be used as a drafting tool:

> As with PROLOG, however, we were able to derive a limited number of more general consequences of the act. This ability is potentially quite important. It means that an executable, logic-based representation of rules and regulations can be used not only to apply the rules, but to aid the process of drafting and redrafting the rules in the first place . . . A similar observation was brought to our attention when we first demonstrated our implementation in January 1984 to officials from the Home Office who were involved in drafting the act (ibid.: 371);

and secondly, it can be used to give advice: 'An obvious application of the formalization of the act is to determine whether in a given circumstance a particular individual is or is not a British citizen.' . . . (ibid.: 377).

It is not the place here to criticise the BNA project (that has been done elsewhere — Leith, 1986); we shall simply list some aspects of the project which are of interest to the general argument in this chapter. However, it is useful to note that this BNA project has been perceived favourably by those in the computing world. The evidence is extremely anecdotal but, as someone working in the general field, there seems to be a general feeling that Imperial 'is setting the pace'. Most certainly, the project is being taken seriously by academics and government departments.

As to an overview of the project we might note several points about it. Firstly, those aspects of law which might be problematical are omitted (the BNA was chosen because it did not have problematical case law associated with it). Secondly, it is a project which manages to leave behind the controversies which surrounded the implementation of the Act. Thirdly, the legislation is treated as being formalisable — i.e. that there *is* some way in which it can be described as logic. Fourthly, there is no relationship made between the project and traditional jurisprudence. Fifthly, it provides a method of stating who can be a British citizen — there will be no argument over its conclusions. Finally, problems met during the formalisation are assumed to be soluble within the logic programming framework (if need be by extending that logic, not by replacing it).

Generally, we might suggest that the project is rather artificial with regard to law. There is no mention of courts, judges or the problems (related in Renton, 1975) of the drafting of legislation. It is a project which is perhaps more concerned with logic programming than with law; and yet it is a project which seems to offer a method for solving many of our legal problems. It offers a logical law, rather than the confusion of case reports, contradictory legislation and suchlike. It caters for our desire for legal certainty.

5.3 METHOD — THE PROTO-IDEA

It would be possible to say that the commonality between Ramist and Kowalskian logic is the similar conceptual framework within which a logical approach to the law has been taken. Yet it can be argued that there is a more fundamental and important commonality which itself generated that application of the logical techniques to the law: that point in common is the use of *method*.

The concept of 'method' was not one which was clearly delineated in the pre-Ramist mind; and we can see that Ramus offered us just such a novelty. Ramism was a method whereby

245

the general world could be analysed and understood and quantified through use of a core set of rules or, to use current terminology, heuristics. As Miller stated, only a few simple steps were required to understand the world according to Ramist epistemology. The concept of 'method' is not one which is utilised by Kowalski — but nonetheless, that is what he also offers us. He provides us with a method to allow *information to be held and transmitted.* By simply collating the facts of the world into a logic program, we can then generate and handle information. The Kowalskian epistemology allows us not only to understand the world but — just like the Ramist emphasis on teaching and communicating — to pass this information on in a highly useful way and, sometimes, to draw new inferences. Both Ramus and Kowalski offer us the same epistemological vision; one utilised the new technique of printing whilst the other utilises the new technique of computing, yet under that technological surface their goals are identical. What we have here is a second, and more fundamental proto-idea, that of 'method'.

The proto-idea of the logic/law thesis has been found in intellectual history for a substantial period of time: yet most proponents of this view have seen the relationship as being as much a psychological one as being one which deals purely with clarity of expression; it is the logic programming camp and the Ramist camp who have set aside the psychological issues and concentrated purely upon the informational issues, upon the transmission of data. Note the similarity between the purpose of Fraunce's thesis and that of PROLOG's legal thesis:

> The telling point in Fraunce's defense, however is that . . . he is concerned about equating teaching with logic, rather than with speech or communication — although he implies the latter equation. Logic does not govern thought, as we should suppose today (man's intellectual processes related to their object), but more specifically does govern communication (the verbalized relationship of man's thought to its object in the presence of another individual), which itself will be resolved into teaching. Furthermore, it must be emphasized that (. . .) Fraunce, and the hundreds like him are not writing in terms of their own insights into linguistic or logic; they are writing as teachers, explaining to their boys what is essentially a textbook tradition (Ong 1958: 159).

Now, we have no evidence that Kowalski believes that logic describes thought: what we do have is evidence that he believes that logic describes *information*. Indeed, it is not too large a step from seeing Fraunce's 'communication' resolved into the contemporary notion of 'information', which we find in multitudes of late twentieth-century thought — 'information technology' and the 'information age'. The importance of Kowalski's method here is that he does not concern himself with the psychological issues, only with the communicational issues:

> There is only one language suitable for representing information — whether declarative or procedural — and that is first-order predicate logic. There is only one intelligent way to process information — and that is by applying deductive inference methods (Kowalski 1980: 44).[4]

Those others in the field of logic and law have, almost without exception, seen a psychological element in the relationship. Take the example of Boole's use of Jewish dietary laws in his formative 'Laws of thought', or take those in the current area of deontic logic who argue over how best to represent the concept of 'legal obligation' which is, of course, a human value and not an attribute of abstract information. Kowalski's method frees us from such problematical areas: logic becomes a cleaner and more technical process for the representation of information.

Ramist epistemology was much criticised in its time, for one of the problems of his method was that it was generalised before the practical and empirical application details were considered — for, as we saw, the method preceded the application in a variety of disciplines (mathematics, theology, etc.). Ramist logic thus constrained the Ramist perspective. We have mentioned how the logic programming researchers have omitted discussion of law or jurisprudence from their work; and yet they have brought their logic programming epistemology to work in the legal domain. It is an attitude which is, as Turkle might suggest, strikingly imperialistic. A critique of legal logic programming (Leith, 1986) has attacked them on just this front, suggesting that they have ignored much of the practical and pragmatic elements of the law — that their epistemology has hidden from them a more subtle interpretation of law and its relation to legislation. We have not yet seen how this critique will be met; but it might be expected that it will be met in the same manner as the suggestion that logic cannot give answers to all mathematical questions. Take Kowalski's response to one such point, that logical inference cannot explain how we know that the square of 5 is greater than the square of 3 simply from 5 being greater than 3:

I find your example from arithmetic puzzling. Given $5 > 3$ to conclude that $5^{**}2 > 3^{**}2$ we need logic + the knowledge that

$x^{**}2 > y^{**}2$ if $x > y$

Of course, logic is not everything. It needs subject matter to work on; and it can be argued that the subject matter is even more important than the general logic itself. (Kowalski, personal communication, 1984).

So logic becomes divorced from knowledge, except as a means of explicating that knowledge. Logic acts, epistemologically, upon facts which arise elsewhere. Nowhere, to my knowledge, has Kowalski written of just where this knowledge arises, nor whether *he* sees such non-logical aspects being more important than the logical; such omissions allow logic programming to set aside any questions about the relationship of logic to knowledge, in the same way that Fraunce could set aside discussion of the role of logic *in* legal knowledge: for logic to Fraunce was a way of teaching and communicating. Logic is thus seen as a medium of legal communication and not legal reasoning. The method (whether of Ramus or Kowalski) stands apart from psychology.

The problem which has continually been met by those who use logic is that it is much easier to use it for the analysis of the world than for the genesis of new information about the world. One reason for this is that logical description involves an inward looking at the subject — it is dissected and quantified into parts. When we place a universal or existential quantifier in front of a logical expression, we delimit our logical world and say, in effect, we are not interested in what lies beyond those borders. The continual quantification thus makes us much more particular and pedantic — we cannot, by the very nature of the limitations which we place upon our described world, make leaps of synthesis. Logicians become pedants: butterfly collectors of the intellectual world. In many ways it is this butterfly collecting from which both Ramus and Kowalski broke free: their respective methods are outward looking, breaking the barriers of analysis and developing into genesis. We can see how Kowalski *et al.* have seen this genesis being developed in the legal world:

> We believe that many of the potential advantages of representing rules and regulations in computer-executable logical form are independent of the actual use of computers. Representation in logical form helps us to identify and eliminate unintended ambiguity and imprecision. It helps clarify and simplify the natural language statement of the rules themselves. *It can also help to derive logical consequences of the rules and therefore test them before they are put into force* . . . (Sergot *et al.* 1986: 371, my emphasis).

Of course, we have no evidence that this technique of deriving legal/logical consequences will work — that we can use logical techniques to discover logical consequences of an act. However, that is inconsequential to the Kowalskian methodology — genesis follows synthesis which follows programming in PROLOG. The methods, then, offer a means of overcoming the previous limitations of logic. Perhaps this is one of the reasons why PROLOG is an idea which has, perhaps accidentally, met its time — for there has always been in existence the proto-idea that logic clarifies. However, clarification is only one part of the necessary equation for usefulness: the other part is that we can use the clarification in a practical way. PROLOG, its proponents argue, offers just this in a method which everybody can use (for example, the legal rules relating to the British Nationality Act project were first extracted from the legislation by a *student of computer science*: not even a student of law is required).

Given the existence of 'method' over a period of 400 years or so, what of the social conditions which might have been in existence in both Renaissance times and in these times which have given rise to logic programming/Ramist dialectic. Might there be a structural aspect of society which has thrown up our two distinct and yet similar methods? It is Elias's contention that post-medieval intellectual life was characterised by a 'loosening of the social bonds'. Note the comments he makes upon the writings of another humanist of that time:

Erasmus's treatise comes at a time of social regrouping. It is the expression of the fruitful transitional period after the loosening of the medieval social hierarchy and before the stabilizing of the modern one. It belongs to the phase in which the old, feudal knight's nobility was still in decline, while the new aristocracy of the absolutist courts was still in the process of formation. This situation gave, among others, the representatives of a small, secular-bourgeois intellectual class, the humanists, and thus Erasmus, not only an opportunity to rise in social station, to gain renown and authority, but also a possibility of candor and detachment that was not present to the same degree either before or afterward. This chance of distancing themselves, which permitted individual representatives of the intellectual class to identify totally and unconditionally with none of the social groups of their world (. . .). (Elias 1978: 73).

It seems hard to believe that the social conditions which gave rise to humanism, Ramism and Fraunce's *The Lawyer's Logicke* are quite being met by the changing social conditions which have given rise to the computer technocrat and the AI boffin. The question which immediately prompts itself is: have we seen such methodologies arise in other periods of techno-social change? Of course, it might well be that it is an aspect of all technocratic groups which attempt to push themselves into a higher social position, that they will use their technical worldviews as 'method' to carry out imperialistic invasions. If this is so it might be easier to explain the mechanisms whereby their thought circle can carry out an attempted invasion of that other discipline, than to explain just why they use 'method' to do it. The latter explanation might simply be that since Renaissance times, the thought style has allowed 'method' to be used as such a tool; but it might be much more complicated.

Imperialism in academic circles is, in the early stages of a discipline, easily carried out; for as long as the invading thought circle does not trespass upon the questions which have been appropriated by the host discipline, little inter-disciplinary communication will arise (and as Fleck has noted, experts in one field will tend to expect other experts from other disciplines to have concrete knowledge of their own field — thus, legal logic programming has been accorded no ill-will nor much attention from the typical academic lawyer). We can see this imperialism in legal logic programming, where those questions which have (over and over again) been taken up, discussed and argued over by the legal/jurisprudential thought circle, have been ignored by the logic programmers — for the logic programmers have concentrated upon technical issues.

What might happen, however, if a PROLOG legal program is implemented (by, say, the DHSS if the current Alvey-related project is considered successful) and it is seen as actually giving 'legal advice' on the law relating to welfare rights. Might the legal establishment not expect to discuss this PROLOG project in their own terms, and see it in an other than technical light? At that point might we expect to see, though on an appropriate scale, a re-run of the battle between the arts scholastics and the humanists.

5.4 CONCLUSION: PROTO-IDEAS, INVOLVEMENT AND DETACHMENT

Elias has argued that the shift from an involved position to a detached position is one which marks the 'scientific' from the 'non-scientific'. He is concerned with the questions which problems in the social sciences pose for those who attempt to investigate them with a 'scientific' spirit, often transferring models of scientific investigation (from the physical sciences) across to fundamentally different problem types. It seems to me that *method* is just such a misplaced transfer: it results in a form of pseudo-detachment where the method is perceived as something which exists both outside of the individual and outside of society: it is a part of the framework of the world. Yet it is a framework which arises from a highly internalised and involved belief in the ability of the world to subsume itself within the operation of that 'method'.

There is a strong element of belief in the power of logic programming, just as there was a strong element of belief in Ramist logic. The world is seen to be a logical world which can be described and manipulated by both methodological systems: yet if this were to be a detached position (say, of logic programming in the law) we might expect the claims made to be less circumspect and more measured. We do not see this: we see instead that the method comes before the application, that there is a belief that the method is appropriate for the application even before the application is more fully examined. When mismatches between law and logic programming are met, we are told that the method requires refining and not that it is inappropriate. Lehnert (1985) has pointed to this as part of the general 'PROLOG problem': for instance, one specific problem for PROLOG (Bowen and Kowalski, 1982) is set aside as requiring a 'smart interpreter', and everything will be well 'just as soon as we figure (how to build it) out.'

A commitment made to a method by a group will continue to make that group try to prove that method. We would expect this, but that does not explain why that commitment was made in the first place. We might suggest it to be a conjunction of today's perspective on 'information' and 'information society' which moulds the thought style of those working in computing and of that long-lasting idea that somehow logic has something to do with the world. These two aspects at least do have a counter in Fraunce's work — the information which was being handled then was the teaching of young teenage MA students, and logic had also imprinted itself deep upon the intellectual vision of the world (and little wonder, since it was so emphasised in the educational process). Whatever the socio-cognitive origins of these beliefs — in logic programming and in Ramist logic — they seem to incorporate elements of the same proto-ideas: the logic/information and method ideas. This reading, though, causes problems for a Fleckian view of the nature of a proto-idea, for he seems to suggest that by their very nature, they are from the pre-scientific past. Yet in the logic programming case, the past can be clearly seen in the present; so can we therefore speak of Ramism as a proto-idea, or is it something else? Fleck gives us little indication; perhaps it is because science has had such a speed of development over the past few centuries that ideas in that domain are quickly turned over, discarded or developed into more useful conceptions by the changing thought style. Of course, we can see that Fleck is indeed correct in viewing proto-ideas as historical elements, but only because proto-ideas to Fleck relate to the scientific esoteric circle. They are not relevant either to the exoteric circle, or to the world at large except as ideas which have been discarded by the esoteric workers.

If we examine examples of proto-ideas which Fleck gives us for the current conception of syphilis, we can see that there is a similarity with Elias's notion of the spectrum of involvement and detachment. Fleck points to 'befouled blood', astrological causes, punishment for sinful lust, mercury as a 'natural' remedy. All these have a highly involved aspect, putting to the centre of the stage the human condition. Moreover, as the science of syphilology developed, these involved aspects became less evident, and the detached aspects became more evident — the fact of the disease became detectable by a detached chemical test. Yet the notion of 'befouled blood' has not — as Fleck would agree — left the stage entirely: think of the herbalist's adverts for tablets and potions to 'purify the blood'. Also, astrology can be readily found and AIDS is seen by some as punishment for sinful lust. The actual problem is that the notion of proto-idea cannot be separated from the notion of esoteric scientific circles: an idea can only be a proto-idea if it is linked to one particular esoteric viewpoint. This leads us to arguing over whether, say, logic programming is 'scientific' or not. If it is, then it can have proto-ideas; if not, then it cannot, by definition, have them. (Or alternatively, we find some way of using proto-ideas as the criterion for what is or is not scientific — and argue further over the grey areas such as logic programming.) The notion of proto-idea thus gives the sense of their being a black and white division between the scientific and the non-scientific; Elias's involved/detached continuum does not.

We can draw together the two theoretical positions and state that the notion of 'proto-idea' is but another way of referring to 'involvement', and that the reason why proto-ideas in the social sciences have altered little (i.e. there is no distinction between many proto-ideas and ideas) is because the social sciences themselves, relative to the hard sciences, have achieved little general detachment. We can, of course, argue over where the borders of any particular definition are to be drawn. However, it seems to me to be much more useful to view abstractions as windows upon the world — each window looking inwards at a different angle, letting light in on different aspects of the epistemological subject. Thus, with the notions of proto-ideas and involvement, the first is suitable only for the viewing of scientific ideas (and, really, providing little evidence of their proto-origins, except through casuistry, as Fleck himself noted); while the second seems to offer a means of talking about the application of science to questions which are typically the domain of social sciences, or even of computer science.

Thus, for example, we can view the development of computer programming languages as the attempt to move away from instructing 'the machine' in terms of, say, storage manipulations (as machine language or assembly-level coding can be most easily interpreted) towards more human-orientated representations. The problem then, for computing as a science, lies in becoming more and more detached in deciding just what 'human-orientated' actually means. Kowalski's logic programming stands at one extreme of the involved/detached continuum: it is not some pre-scientific or social science project which has little relationship to computer *science*. It is but one example of the design of a programming language which reflects the way we need to think of real-world programming problems. Other examples (which might be less detached) are object-orientated programming systems, or even those which use pictorial representations in place of the 'system commands' to the operating system. Unless we realise this general notion — that programming languages frequently represent our view of what the world actually is — we can never fully appreciate the object of computer science, nor of AI.

Studies of AI which concentrate upon the machine, or upon the distant goals, forget that most of the day-to-day effort of AI is concerned with the mundane use of programming languages: the programming effort which goes into any decent-sized program (say, to contribute towards a PhD) can be calculated in the thousands of hours. It is little wonder that languages — and their relationship to data structures — can become the centre of much argument: look to the never-ending discussions in the semi-popular as well as academic literature over the merits and demerits of PASCAL, BASIC, ADA, MODULA, LISP, FORTH, LOGO, etc. When a language is used to model the problems of the world (whether problems of data handling for gas billing, or for problems of representing models of mind), we must expect a strained relationship between the need to detach, and the desire to involve. The belief in PROLOG is a case where the desire to involve has won over the need to detach.

Elias himself pointed out that the notions of involvement and detachment would be of little use if they were seen to be two parts of a dichotomy — if they were to 'adumbrate a sharp division between two independent sets of phenomena' (Elias 1956: 227). The problem with 'proto-idea' is that it *does* indicate an independent set of phenomena: those which are, to the esoteric circle, involved. In terms of usefulness, therefore, perhaps we should discard the notion of proto-idea and more fully develop the notion of involvement and detachment. This might enable us to analyse more clearly the role of programming languages in the field of AI.

NOTES

1. Of course, this debate on just what 'logic was' was limited to a section of the intellectual elite: we should not forget that in the wider world some scepticism to any logic from both scholastics and anti-scholastics was met — best evidenced, perhaps, by the sixteenth-century proverb 'Logic, Law and Switzers can all be hired to fight on one's side.' The Swiss were, at that time, Europe's mercenaries.

2. Before moving on from Lighthill, it should be noted that a more 'mainline' (if that is the correct expression) computer scientist, Needham (who supported Lighthill in the report), pointed out that many of the claimed inventions of AI in programming technology were not that discipline's inventions. Even LISP was slighted: 'list-processing is a technique for burying store-management problems, excellent for rich people with complicated programs to write' (Needham, 1973).

3. We are interested in the general perspectives taken at Imperial; lately, one or two of the researchers have broadened their perspectives, yet stayed relatively firmly aligned to logic programming. This broadening to discuss some aspects of jurisprudence is recent and quite limited: for present purposes, we will ignore it.

4. This statement was made in reply to a questionnaire published in *SIGART* as a 'special issue on knowledge representation'. Unfortunately, Kowalski only replied with a position statement — the answers to the 49 questions in that questionnaire might well have elucidated Kowalski's position more clearly: one began, 'Suppose that you were appointed philosopher-king . . .'.

REFERENCES

Balbin, I. and Lecot, K. (1985) *Logic programming: a classified bibliography*. Wildgrass Books, Victoria, Australia

Bowen, K. and Kowalski, R. A. (1982) Amalgamating language and metalanguage in logic programming. In K.L. Clark and S-A. Tarnlund (eds), *Logic programming*, Academic Press, London, pp. 153–72

Elias, N. (1956) Problems of involvement and detachment. *British Journal of Sociology*, 7, 226–52

——— (1978) *The history of manners: the civilizing process*, vol. 1. Basil Blackwell, Oxford

Fleck, J. (1982/1987) Development and establishment in Artificial

Intelligence. In N. Elias, H. Martins and R. Whitley (eds), *Scientific establishment and hierarchies*, Sociology of the sciences, vol. VI, Reidel, Dordrecht, pp. 169–207; reprinted as Chapter 3 in this volume

Fleck, L. (1979) *Genesis and development of a scientific fact*. University of Chicago Press, Chicago

Fraunce, A. (1588) *The Lawyer's Logicke*. Facsimile printed by the Scolar Press, London

Guest, A.G. (1961) Logic in the law. In Guest (ed.), *Oxford essays in jurisprudence*, Oxford University Press, Oxford, pp. 176–97

Hammond, P. (1983) 'A listing of a Prolog program describing entitlement to supplementary benefit'. Department of Computing, Imperial College, University of London

Henry, D.P. (1972) *Medieval logic and metaphysics*. Hutchison, London

Hogger, C.J. (1984) *Introduction to logic programming*. Academic Press, New York

Kowalski, R. (1970) 'Studies in the completeness and efficiency of theorem proving by resolution'. Unpublished PhD thesis, University of Edinburgh, Edinburgh

———— (1974) Predicate logic as a programming language. In *Proceedings IFIP-74 Congress*, North-Holland, New York, pp. 569–74.

———— (1979) *Logic for problem solving*, North-Holland/Elsevier, New York

———— (1980) Response to questionnnaire published in *SIGART Newsletter, 70*, February (Special issue on knowledge representation)

———— (1981) Logic as the Fifth Generation computer language. In *The Fifth Generation?*, Infotech Computer State of the Art Reports, Pergamon, Oxford, pp. 76–87

———— (1982) Logic as a computer language. In K.L. Clark and S-A. Tarnlund (eds), *Logic programming*, Academic Press, London, pp. 3–16

Lehnert, W.G. (1985) The Prolog problem. *The Journal for the Integrated Study of Artificial Intelligence, Cognitive Science and Applied Epistemology (CC-AI), 2(4)*, 3–11

Leith, P. (1986) Fundamental errors in legal logic programming. *The Computer Journal, 29(6)*, 545–54

Lighthill, Sir J. (1973) Artificial Intelligence: a general survey. In *Artificial Intelligence: a paper symposium*, Science Research Council, London, pp. 1–21

MacCormick, D.N. (1979) *Legal reasoning and legal theory*. Oxford University Press, Oxford

Michie, D. (1973) Comment. In *Artificial Intelligence: a paper symposium*, Science Research Council, London, pp. 38–45

Miller, P. (1961) *The New England mind: the seventeenth century*, vol. 1. Beacon Press, Boston

Needham, R.M. (1973) Comment. In *Artificial Intelligence: a paper symposium*, Science Research Council, London, pp. 32–4

Ong, W.J. (1958) *Ramus, method, and the decay of dialogue*. Harvard

University Press, Harvard

Raphael, B. (1976) *The thinking computer: mind inside matter*. Freeman and Company, San Francisco

Renton, D. (1975) *The preparation of legislation: report of a committee appointed by the Lord President of the Council*. Her Majesty's Stationery Office, Cmnd. 6053, London

Sergot, M.J., Sadri, F., Kowalski R.A., Kriwaczek F., Hammond P. and Cory, H.T. (1986) The British Nationality Act as a logic program. *Communications of the ACM, 29(5)*, pp. 310–86

Turkle, S. (1984) *The second self: computers and the human spirit*. Granada, London

Warren, D.H.D. (1981) A view of the Fifth Generation and its impact. In *The Fifth Generation?*, Infotech Computer State of the Art Reports, Pergamon, Oxford, pp. 146–60

6

Expert Systems, Artificial Intelligence and the Behavioural Co-ordinates of Skill*

H.M. Collins

6.1 INTRODUCTION

In our lives we are going to encounter more machines that are more 'intelligent'. In this chapter I will try to explain what this means and how we should think about it. My starting point is ideas drawn from recent interpretations of Wittgenstein (e.g. Winch, 1958; Bloor, 1983) and the sociology of scientific knowledge (e.g. Collins, 1985).[1] I intend to use these ideas to look at the development of intelligent programs called 'expert systems'. Expert systems are with us now. They are relatively accessible in the sense that the principle underlying their design is easy to grasp, and in that they are not too difficult to build. I think that the ideas underlying expert systems are essentially the same as the ideas underlying much work in Artificial Intelligence (AI). Even if there are new or different principles of AI that are not encompassed by the model advanced here, I hope to show (implicitly) the kind of difference in design principle that will be required to make an 'in principle' in the ability of machines to think like humans. It will not be a difference in quantity.

The chapter starts with a simple description of expert systems and then sets up a model of human culture intended to bridge the gap between the way the 'knowledge engineers' think and the way sociologists/philosophers of knowledge think. Using this model, I go on to develop a four-stage classification of expert systems, with predictions of the likely success of the different types. The chapter concludes by setting the discussion in the context of the larger debate about the possibility of AI, and shows how intelligent machines may come to seem to think like us, and how this illusion might be dispelled. The crucial ideas in the chapter are the inescapable input of an end user in an intelligent interaction, and the notion of 'behavioural co-ordinates of skill and social action'.

An expert system is a computer program. It differs from other programs in that it encodes relatively familiar know-ledge expressed in non-mathematical form. The programming languages of many expert systems are designed to reflect the ordinariness of the knowledge base, and programs can be comparatively easily read by the non-expert. Compare Figure 6.1, a simple BASIC program for generating and analysing a list of random numbers, with Figure 6.2, an extract from an expert system for semi-conductor crystal growers written with 'AUG-MENTED PROLOG FOR EXPERT SYSTEMS' — (APES) (Collins, Green and Draper, 1985).

Figure 6.1: A program in BASIC

```
20 RANDOMIZE
30 FOR I=1 TO 500
40 A=RND * 5000
50 PRINT USING "££££";A
55 IF A <100 THEN P=P+1
60 IF A <1000 THEN M=M+1
70 IF A >4000 THEN N=N+1
75 IF A <10 THEN Q=Q+1
80 NEXT I
90 PRINT "numbers less than 1000 =" M
100 PRINT "numbers more than 4000 =" N
110 PRINT "numbers less than 100 =" P
120 PRINT "numbers less than 10=" Q
```

The specialist programming languages, such as that in which the second program extract is written, can make decoding and alteration of expert systems easier and they may also make it easy to query the knowledge base of the system. There is no reason why such systems have to be about technical subjects: they can easily encapsulate 'rules of thumb' (heuristics, in the jargon) which would not normally be formalised in this way. For the purposes of illustration, and demystification, I wrote a very small program in APES called 'PICKUP'. The complete program is represented in Figure 6.3. It could be thought of as a kind of miniature expert system.

Figure 6.2: Extract from an expert system for crystal growers

---

(X:Y) wont-melt-container-material Z if
    (X:Y) has-melting-point x and
    Z will-soften-at y and
    x is less than y

X wont-be-dissolved-by-or-reduce (Y:Z) if
    not (Y:Z) dissolves-container-material X and
    not (Y:Z) will-be-reduced-by X

(X:Y) dissolves-container-material Z if
    (X:Y) contains-metal and
    Z is-a-metal

(X:Y) will-be-reduced-by Z if
    (X:Y) contains-an-oxide and
    Z is-a-reducing-agent

pyrex will-soften-at 501
quartz will-soften-at 1251
platinum will-soften-at 1900

Figure 6.3: A simple expert system: 'PICKUP'

X will-go-back-to-my-place if
    X has-a-fleeting-smile and
    X will-maintain-prolonged-eye-contact and not
    X is-short-sighted

An actual consultation with PICKUP is represented below.
(Read ':' as 'such that'.)

[User]	find (X:X will-go-back-to-my-place)
[Computer]	which (X:X has-a-fleeting-smile)
[Computer]	Answer is
[User]	Mary
[Computer]	Answer is
[User]	Delilah
[Computer]	Answer is
[User]	Albert
[Computer]	Answer is
[User]	end

[Computer]   Is it true that Mary will-maintain-prolonged-eye-contact?
[User]   no
[Computer]   Is is true that Delilah will-maintain-prolonged-eye-contact?
[User]   yes
[Computer]   Is it true that Delilah is-short-sighted?
[User]   yes
[Computer]   Is it true that Albert will-maintain-prolonged-eye-contact?
[User]   yes
[Computer]   Is it true that Albert is-short-sighted?
[User]   no

[Computer]   ==> Albert

As can be seen, the reasoning works in a kind of 'tree' structure. From the initial query, routes through the ramifying branches of the tree are explored, touching here and there on the user's knowledge until a satisfactory answer that makes sense in the enquirer's universe is reached. However, the question of how and where the tree touches the user's universe is a vexed one.

In some expert systems (such as those written in APES), the program makes it possible to save the information 'learned' at each consultation. If the above consultation had been saved, then the next time PICKUP were asked the same question, it would immediately conclude with the answer 'Albert' without the intervening queries. (Of course, this might not always be a good thing!)

The specialist languages, useful though they are, are not the essence of expert systems. In principle, any computer program can be written in any computer language, so the second and third programs listed above *could* have been written in BASIC, albeit with more difficulty, and both languages amount to strings of '0's and '1's when they are translated for action in the processing heart of the computer. There is no agreed definition of an expert system: some think that the essence is the language, others that a probabilistic inferencing mechanism is the vital ingredient, and so forth. The definition I will adopt is one that delimits the set of programs that is particularly interesting from the point of the view of the sociologist of knowledge, or more generally, the 'knowledge scientist'. (Those who actually build expert systems call themselves 'knowledge *engineers*'!) I define an expert system as 'a computer program designed to encapsulate the knowledge of experts'.

There are two phenomena that make these programs interesting from the knowledge scientist's point of view. Both arise out of the whole AI debate, but expert systems make the problem more immediate and tractable. The first point is that any attempt to articulate knowledge — and it must be articulated before it can be encoded — sheds light on the degree to which knowledge and culture can be articulated and transferred by non-human media.[2] This is a question which touches upon the deepest concerns of anthropologists, philosophers, historians and sociologists, particularly those of the interpretative persuasion. The second point of interest is that the extent to which skills can be encoded is the extent to which machines can replace skilled craftsmen in the workplace and this is of enormous significance to sociologists and historians of industrialisation as well as to anyone concerned with the relationship between humankind and machines.[3]

The sociologist's crucial input to these questions (and I owe this to Oldman and Ducker, 1984 — see also Suchman, 1985) is that the user of an 'intelligent' machine can make up for a great deal of its deficiencies. Oldman and Ducker point out that Garfinkel's (1967) 'counsellor' (where students were effectively 'counselled' by a list of random numbers) reveals the extent to which recipients can interpret creatively in order to make sense of minimal information. Collins (see for example Collins *et al.*, 1985) has argued that expert systems cannot encode 'tacit knowledge' but that this is not necessarily disabling because 'end-users' can provide what the program cannot. The implication is that the expert systems will only be able to be used by those who can fill in these gaps — that is, by those who are fairly skilled to start off with. Thus, the future of expert systems lies in their role as assistants to the skilled rather than as replacements for the skilled. The typical end-user, say for a legal expert system, will not be a novice; a skilled lawyer will still be needed to interpret and make judgements about the validity of the system's decisions (see also Leith, 1986). In circumstances where the end-user is not skilled, the system will either be unable to put the user in a position to do more than before, or it will provide potentially misleading advice (Collins *et al.*, 1985, Dreyfus and Dreyfus, 1986).

To summarise this point of view, one might say that the standard model of expert systems, and many other knowledge transmission media, is one in which all the knowledge flows from the expert, to the user, or recipient, via the medium, thus:

EXPERT- - - - - - - - ->MEDIUM- - - - - - - - ->USER

The sociologist's view is that since that which can be encoded within the medium must be interpreted before it can make any sense, expert and user must share a framework of interpretation — a culture, or 'form-of-life' — if knowledge is to pass in this 'left-to-right' direction. To express this another way, the user needs to contribute something in a 'right-to-left' direction before the left-to-right flow can take place. In extreme cases such as Garfinkel's counsellor, the user 'invented' the expert at the left-hand end and this, we could say, amounted to right-to-left 'knowledge transfer' without any left-to-right component at all. In the normal way, however, the user does not need to provide so much.

The model just described is rather mechanical; hermeneuticians would argue that the message obtained at the right-hand end of this chain is not the same as the message that went into the left-hand end. They would say that the eventual message is something new, an amalgam of the expert's input and the user's interpretative framework. I imagine that phenomenologists would say something similar — in terms of structures of relevance — and that ethnomethodologists would be more concerned with the way users make, or fail to make, orderly sense out of what they received rather than the content of the message (Suchman, 1985; Woolgar, 1985). With expert systems, however, additional obstinate and pressing questions remain; 'Could an expert system capture laser-building knowledge and transfer it better than written acounts?'; 'Can a skill be encoded in a system sufficiently well for it to be transferred to future generations, even though all human practitioners have died?'; 'Can apprentice-masters be replaced with machines?'.

6.2 THE RULES MODEL

To consider these questions we must have a way of describing the cultures which experts and users of knowledge share, and the description must mesh with our way of talking about computer programs. For these purposes I propose to adopt what I will call 'the rules model' of culture — that is, I intend to speak of cultures as comprising sets of rules (Collins, 1987). This is similar to the model used (perhaps implicitly) by many of those working in AI and expert systems, though they do not necessarily think of themselves as modelling 'culture'. However, the model I adopt needs to be qualified in a number of ways at the outset. Because, as Wittgenstein pointed out, rules do not contain the rules of their own application, sets of rules which model cultures would be infinite in extent were we to try to explicate them in such a way as to cope with the open systems in which human cultures exist. There are many examples of the problems which arise when humans attempt to follow rules as though they were algorithmic in nature: for example, one way of creating *trouble* is to 'work to rule'. Alternatively, consider this clipping from the *Arizona Daily Star* of Saturday, 31 May, 1986.[4]

> CHICAGO - A rookie bus driver, suspended for failing to follow correct emergency procedures when a girl suffered a heart attack on his bus, was following overly strict rules that prohibit drivers from leaving their routes without permission, a union official said yesterday.
>
> 'If the blame has to be put anywhere, put it on the rules that those people have to follow' [said the union official].
>
> [A transit authority spokesman defended the rules. He said] 'You give them a little leeway, and where does it end up?'
>
> [The driver] will be suspended for three days, will receive instructions on the rules and procedures during that period and will have to 'demonstrate an understanding of those rules to be reinstated' . . .

The problem here is that training the bus driver in the meaning of the rules in a formalistic way requires that every eventuality that might affect his driving during the whole of his life be foreseen! How the driver would clearly *demonstrate* that he had such an understanding is not at all obvious.

Even highly formalised sets of rules turn out to be less than straightforward when operated in the open system of life: for example, these two incidents took place on the same day, 21 June 1986. During the final penalty 'shoot-out' in the World Cup football quarter-final between Brazil and France one of the French players struck the post with his penalty kick and the ball bounced back. The ball hit the Brazilian goal keeper, who was a yard 'off his line', and rebounded into the goal. There is no doubt that if it had hit a French player and rebounded it would not have been counted a goal, for the ball would have been 'dead' after it bounced back from the post. However, the rules, it seems, did not cater for this other unforeseen eventuality. The referee awarded a goal.

Earlier on the same day, in the cricket Test Match between England and India, the Indian 'twelfth man' remained on the field by accident while an Indian player bowled a complete over of six balls. What would have happened if a batsman had been out during this over? One commentator suggested that a batsman could not be out under these circumstances since the ball was 'dead' during the entire over; the game of cricket is played between two sides of eleven men, and in this case one side had twelve men. However, on this interpretation, what should have happened to the runs that England scored during the over? Also, was an over actually bowled, or did the same bowler effectively bowl twelve balls from the same end since the intervening over did not count? Did India bowl the requisite number of overs required of them in the day? The balls could not have been treated as 'no-balls' or the over should have been six balls longer and England should have been given six extra runs. In any case, a run out — which is allowed off a no-ball — would not have been just under these circumstances.

In these two cases of very well established sports, with highly codified constitutive rules, played at the highest standard and with the careful planning that befits world-class events, the rules were not adequate in themselves to cope with unforeseen circumstances. In such circumstances we usually think of the rule base as *deficient*; we think that the rule base is potentially perfectible given enough time, experience and dedication. However, this is to put the cart before the horse. What the examples really show is that all attempts to formalise human decision-making are inadequate. The reason that formal rules, such as the constitutive rules of cricket and football, work at all is because human agency intervenes to establish their meaning by fiat where their imperfections are exposed. That is why the ultimate rule in these sports is 'the umpire's/referee's decision is final'. That is why the French penalty stands, and why England were defeated by India in the Second Test. It is also why Argentina beat England in the World Cup quarter-final held on 21 June, in spite of the fact that Maradona handled the ball into the net for the first goal. It is not that humans are imperfect, or that formal rules are perfectible beyond the capacities of humans, but that no finite system of rules can cope with every eventuality in an open system and thus, frequently, decisions can only be made with human intercession. The only perfectible set of rules is an infinite set, but this it to say little, for an infinite set of rules — which by definition can cope with any eventuality — is merely a description of the whole past and every possible future of the cultural universe.

In the 'rules-model' that I am putting forward, no complete culture could ever be written out. Indeed, at any point only a small subset of the rules can be articulated. Quite apart from the fact that one cannot articulate an infinite set, new rules are continually being articulated as a result of scientific research.[5] What is more, as we know from studies of science, scientific research does not always lead to consensus, so that the very notion of a fixed set of rules that could be 'discovered' is naïve. Furthermore, the creative possibilities of human culture are not captured by the notion of a set of rules — except that the notion of an infinite set of rules provides for a rule for every contingency including every creative possibility and every negotiating tactic over what counts as a possibility. Thus, the notion of an infinite set of rules has a kind of auto-destructive quality, but this is no bad thing. It makes clear that the rules model is not a realist model as far as psychology or sociology are concerned — there is no future at all in looking for these sets of rules in our heads or in social collectivities;[6] indeed, to state that the sets of rules are infinite in extent, is to say that this cannot be how cultures are expressed in our heads or in our society. Nevertheless, it is a useful model.

The advantage of the rules model is that it makes immediate contact with AI and knowledge engineering; expert systems mostly comprise assemblages of rules. A second advantage is that the model avoids questions about what facets of human knowledge can and cannot be explicated in principle: the model allows that any limited aspect of human knowledge can be explicated as rules in principle, but that the aspect that is explicated can only be used and understood because it rests on the infinite foundation of rules that remain unexplicated but are nevertheless shared by other members of the culture. If this working model *is* accepted, it follows that no computer program will ever be equivalent to a fully socialised human (so long as it is necessary to explicate knowledge before it is encoded). It does not follow, however, that we can pin down specific bits of knowledge that cannot be explicated in principle — though a standard method of arguing about whole spheres of knowledge will be to show that to encapsulate them leads to the familiar exponential explosion in the number of rules required for their capture. The model also allows us to talk about what the user contributes to knowledge transfer — the infinite rule base. Finally, the model allows us to talk about what will happen as machines get bigger and faster so that they can store and use more and more rules; the more rules they can use, the less will be required of the end-user, though since an infinite number of rules would be required to obviate the end-user's contribution entirely, we can see that this progress cannot be extrapolated too far (such an extrapolation would be analogous to committing what Dreyfus and Dreyfus call the 'fallacy of the first step').[7]

6.3 STAGES IN THE DEVELOPMENT OF EXPERT SYSTEMS

We can apply the rules model by using it to generate a simple typology of expert system. From the rules model perspective the simplest type of expert system to build is one in which the program is based on a set of rules which are already encoded in some other form such as a handbook. There is no reason why an expert system should not capture everything in a technical manual of whatever complexity. The system might have a few disadvantages compared with the manual — for example, it is not easy, though it is possible, to read the contents from beginning to end. On the other hand, the system could have great advantages over the book, in that when it is consulted, it can find its way swiftly to the right place in its knowledge base for that particular consultation and provide the information required without presenting large quantities of material that are not immediately relevant. One might say that the system acts as an intelligent 'index' to its data base, tracing just the route through its knowledge that is suited to the particular enquiry in process, combining *all and only* relevant aspects of its knowledge. Expert systems have other advantages that arise from their foundation of ordinary computing capability. As Levy (1987) puts it in his otherwise critical paper:

> The management and accessing of data bases by computers can be more efficient and reliable. Their ruthless logic can be advantageous (and sometimes not!). Their perseverance in search tasks is to be admired. Their memory is beyond that of a herd of elephants. And so on.

Systems that reproduce ready codified information have been built: for example, a system built by British Gas to provide herbicide advice for weed control in dispersed pumping stations was based largely on an existing manual. Expertech, a knowledge engineering firm, is reproducing the British Government's Statutory Sick Pay (SSP) regulations in an expert system for use by employers. In this system, just as in PICKUP, the user will be prompted for a series of replies and the system will thread its way through the relevant regulations until it comes up with the appropriate advice. Expertech's aim is to avoid what a respondent referred to as 'going over the weir'. Their strategy was to build systems (or 'shells' for systems) that would cost a few hundred or a thousand pounds and would do a small, but useful, cost-effective job. 'Going over the weir' amounts to tackling rather more difficult problems involving 'issues of judgement' that could easily cost hundreds of thousands, or even millions of pounds, without any guarantee of a successful product at the end. Expertech tries to design its systems in such a way as to prompt the user to seek further advice outside the computer when questions of judgement arise. My respondent described the current situation regarding SSP in the following terms:

> That, [showing me a 59-page booklet] currently is the DHSS booklet on Statutory Sick Pay. As you'll see, its a nice, readable little booklet. . . . and if you read it from cover to cover three or four times and think about it a lot, you'll gradually understand SSP. . . . a person who has to work

Figure 6.4: The expert fills the gap between machine and user

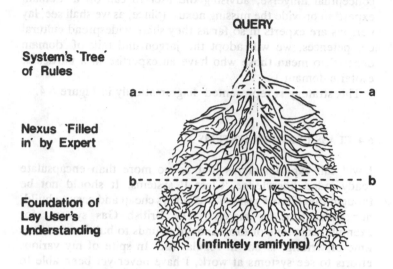

QUERY

System's `Tree`
of Rules

a — a

Nexus `Filled
in` by Expert

b — b

Foundation of
Lay User's
Understanding

(infinitely ramifying)

with SSP, their problem is very much that someone has gone off sick, they then have a problem with next week's payroll, or whatever — Do I have to pay them SSP and how much? . . . And that doesn't give you an inkling. You have to work it out by understanding all the regulations.

Expertech's point was that a system could be built very simply, to be sold for a few hundred pounds, that would solve the employer's problem in most cases, if not in the most difficult ones.

Our knowledge base is not intended to be totally definitive, but there's an awful watershed in some situations where you try to make a knowledge base absolutely definitive in every respect. [Compare this with] something which is pretty helpful in most situations. . . . [we're going for the latter] . . . hopefully building checks in, so that it recognises, or can explain its limits and bounds.

We can summarise this respondent's approach, in terms of the rules model, as making no attempt to encode more rules than may be obtained from ready-coded sources, and whenever this sub-set of rules fails to make contact with the end-user's conceptual universe, advising the user to call on a 'domain expert' to provide the missing nexus. (Since, as we shall see, lay persons are experts in so far as they share widespread cultural competences, we will adopt the jargon and talk of 'domain experts' to mean those who have an expertise in a particular esoteric domain.)

This strategy is represented diagramatically in Figure 6.4.

6.4 CLASS I EXPERT SYSTEMS

I will call expert systems that do no more than encapsulate ready-coded knowledge 'Class I' systems. It should not be thought that Class I systems are always cheap and easy to build, nor guaranteed to succeed. The British Gas system, for example, cost tens of thousands of pounds to build and, at the time of writing, is not in regular use. In spite of my various efforts to see systems at work, I have never yet been able to watch a system of any type being used in an everyday working setting. This does not mean that there are none, and certainly does not mean that there never will be any, but only that they are few and far between.

6.5 CLASS II EXPERT SYSTEMS

A second class of expert system contains a rule base built up largely of 'rules of thumb' or 'heuristics'. These are rules of expertise 'elicited' from domain experts by detailed questioning. The early and widely publicised expert systems used such rules. The famous system MYCIN[8] used medically diagnostic rules, and another celebrated system, PROSPECTOR, was designed to use rules for the interpretation of seismic signals to locate deposits of valuable minerals. A large and growing body of technical literature discusses the problems of eliciting knowledge from domain experts. The knowledge elicitation process is widely quoted as being the 'bottleneck' which limits the widespread adoption of the new technology. Knowledge engineers have found themselves surprised and frustrated at the unwillingness, or apparent inability, of domain experts to express their operating rules clearly or, where necessary, in probabilistic terms.

There is no doubt that Class II systems are much harder and more expensive to make than Class I systems: for example, Karl Wiig of the American financial consultants Arthur D. Little, has suggested that developing a basic system for use in the financial planning environment would take 10 to 20 person/ years, while the deployment of a full system might take 30 to 150 person/years. Obviously the costs involved are to be measured in millions of dollars (but this, of course, is in a section of the economy where lavish resources are most likely to be available.)[9]

In spite of the cost and difficulty, in terms of the rules model the problem is not different in kind from the problem of building systems from ready-encoded rules. This is because so long as the knowledge engineer sticks to rules belonging to the esoteric knowledge base of the domain expert, and so long as these can be articulated, and so long as lay end-users will consult the machine in collaboration with a more expert adviser, so that tacit knowledge is not an issue, then there is no reason why the rule base should explode catastrophically.

6.6 CLASS III EXPERT SYSTEMS

The third class of systems encounters problems of a different order of magnitude. The knowledge base of this class may be founded on the ready-coded rules of Class I or the esoteric heuristics of Class II, but differs in that there is an attempt to do away with the need for interventions of human experts between system and lay user. An example of such a system is that part of the British Department of Health and Social Security (DHSS) 'demonstrator' being built at the University of Surrey by Nigel Gilbert and a team of researchers. Gilbert describes this project as being aimed at

> . . . the general public, the people who have to deal with this huge and complex organisation and its regulations. . . . (We) are building systems to demonstrate how present day computer technology could make claiming benefits easier, and the regulations more comprehensible (Gilbert 1985: 3).

One part of this project is a 'Forms Helper', intended to aid claimants in filling out forms. The DHSS project is based on the ready coded knowledge of the DHSS rule books but, as Gilbert (ibid.) puts it:

> . . . [a person filling in a form] has no context to help decide on matters such as how precise the answer should be, whether an estimate would be acceptable, or whether the claim would be invalid if no answer at all is given to a question. These are all areas where the Forms Helper can assist, with examples of possible answers, with explanations of terms, and with syntactical or even semantic checks on the answers for clarity and consistency.

One might say that the Surrey project, like any other project that aims to deal directly with the lay user, is trying to cope with the Wittgensteinian point that 'rules don't contain the rules for their own application'. Technical experts are needed to show how the rules are to be applied where widespread technical competences (common sense) do not make this clear.

In trying to do away with expert interpreters of the system's output, Class III systems are likely to 'go over the weir', as my respondent from Expertech put it. Thus, the Surrey project is supported by a grant of approximately £700,000, but is intended only to demonstrate what such a system could do, rather than to be an implementable or even a basic system.[10]

We can see how a project of this sort, if it maintained the goal of making the DHSS rules fully comprehensible to the lay person, could run into explosive growth of its rule base. Thus, for example, a standard problem for DHSS claims is the interpretation of the term 'cohabitation', upon which substantial benefit entitlement can depend. If a woman has an intimate friend who works on a North Sea oil-rig for most of the year and therefore visits her only occasionally, is she cohabiting? Such problems would not be dissimilar to the difficulties we can visualise if we cast ourselves in the role of strangers and look at more familiar instances. 'PICKUP', for example, the miniscule 'expert system' described above, only works because its rule base can easily make contact with its users. I would guess that most readers found PICKUP's indifference to the sex of the potential companion mildly amusing. Though the program followed its rule base accurately, it was simply assumed that the user would know that it was necessary to specify the sex of the partner implicitly at the stage when the computer asked for a list of names of those who had 'a fleeting smile'. The user, in the example of interaction set out above, must either have been bisexual, or must not have understood the significance of someone 'coming back to my place'.

We can use PICKUP to illustrate further what an exploding rule base would look like just by expanding it in a literal minded way so that it could be used by someone without Western cultural competence. Under these circumstances further rules would need to be added roughly as follows. I will ignore PROLOG's syntax for this purpose, but all these rules could easily be coded if necessary. (See the Appendix for an example of adding an extra rule to PICKUP and the resulting consultation.)

a smile is fleeting if it lasts for a short time.

a short time means less than one second, but long enough to be clearly recognisable

a smile is a movement of the facial muscles so as to raise the corners of the mouth and the eyes

a smile is not a tic

a tic is an involuntary movement

a voluntary movement can be recognised because it is not repeated regularly and is not made in a jerky manner

a jerky manner implies a sudden start and stop to the action

sudden means in less than about one tenth of a second [it is important for a successful pickup that these time intervals be judged rather than timed — e.g. by stopwatch]

eye contact means contact between pickup target's eyes and user's eyes

eye contact does not mean physical touching of eyeballs

eye contact means that the pupils of the eyes should be aligned in such a way that beams of light passing normally through the user's pupils would converge on one of the target's pupils and vice versa [this should not be tested by experiment]

the pupil upon which the beams would converge should alternate between left and right but not regularly proximity between target and user increases likelihood that eye contact is being achieved except when contact is made 'across a crowded room'

in 'a crowded room' the weighting of eye contact as signifying success is increased

prolonged means . . . etc

By extending the rule base in this way, the program can make sense to a less skilled, and therefore a wider set of potential users.

PICKUP, of course, is an unrealistic example, but the principle remains — as the rule base is widened it can form a nexus between domain expert and wider, less culturally accomplished, groups. What is happening in effect is that the knowledge required by the end-user is being substituted by the rules provided by the system. If we return to Figure 6.4, we could say that the line a——a, which represents the extent of the rules contained in the system, is moving down to meet the line b——b. When they meet, the intervening expert can be dispensed with, and the system can communicate direct to the lay public. This is what must happen if expert systems are to have the future that was promised when they first appeared on the scene (Feigenbaum and McCorduck, 1983; Michie, 1979). Class III systems could act as apprentice-masters, could act as effective media for transferring expert knowledge, and could store expertises that were in danger of dying out. I do not think that we will see Class III expert systems implemented in anything but the most simple areas, where the expertise that needs to be encoded is scarcely more than *information*. 'Information', in this sense, means rules that appear completely transparent because the tacit knowledge on which they rely for their interpretation is already completely and thoroughly distributed in the lay population. (The meaning of information in this sense will, of course, vary from population to population. One culture's information is another culture's expertise! — see Collins, 1987.)

6.7 CLASS IV — EXPERT SYSTEMS AND ARTIFICIAL INTELLIGENCE

If a——a goes down still further — below b——b — the

machine begins to take over functions which are normally the prerogative of the user. Only here do expert systems threaten to become 'intelligent', where intelligent is taken to mean matching the cultural competence of ordinary human beings. Once a——a goes below b——b, we can talk of a fourth class of expert systems. We can represent the development of Class IV systems on another diagram — see Figure 6.5:

Figure 6.5: Expert systems increase in power

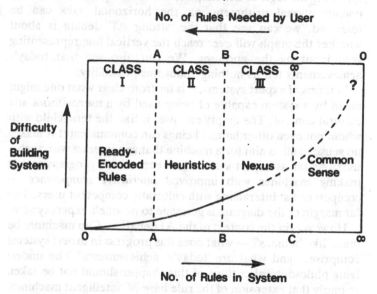

In this diagram, the number of encoded rules in the system increases from left to right (equivalent to the line a——a moving downward in Figure 6.4). The line A——A separates systems which contain ready-encoded rules from systems that need to overcome the 'great bottleneck' of knowledge elicitation in order to encode their heuristics. In terms of the diagram, this is not a particularly significant divide. The line B——B is of greater significance, for here we attempt to do away with an intervening domain expert and this requires that we convert expertise into common sense. At the line C——C the domain expert has been eliminated entirely. This, as I have argued, is not likely to happen unless the domain expert's expertise comprises 'information' alone (for example, a railway timetable expert). As I have argued, it is more and more difficult to make systems as the graph moves to the right. The difficulties begin to increase exponentially around the B——B line. However, as the locus of the curve moves to the right, the cultural competence (number of rules) needed by the user of the system decreases — but the user still needs an infinite number of rules, even at C——C (the system does not necessarily need an infinite number, even at this point). At the right-hand extreme of the diagram, the system contains the infinite set of rules and, following the logic of the chapter, the user requires none. If the unconventional arithmetic on the horizontal axes can be tolerated, we can see that the 'strong AI' debate is about whether the graph will ever reach the vertical line representing zero input by the end-user. We can also see that today's achievements tell us nothing about this possibility.

In terms of expert systems, it is far from clear what one might mean by a system capable of being used by a user without any cultural content. The empty end-user is like the feral child with whom not even other human beings can communicate! It makes no sense at all to aim for a machine that can interact with such a user. As far as expert systems are concerned, progress means making machines with improved *interactive competence* — competence at interacting with culturally competent users. The far margin of the diagram is germane to no-one's expert system.

However, in the context of the AI debate — Can machines be made like humans? — what does this progress in expert systems comprise?; and what *are* 'today's' achievements? The underlying philosophical critique of this chapter should not be taken to imply that extension of the rule base of 'intelligent machines' is not progress; it is progress — progress which looks, to the respectful user, like increasing intelligence. What this progress comprises can be best understood, I believe, by taking a metaphor from Peter Winch's (1958) book, *The Idea of a Social Science*. On page 78, Winch talks of an injured cat:

> We say the cat 'writhes about'. Suppose I describe his very complex movements in purely mechanical terms, using a set of space-time co-ordinates. This is, in a sense, a description of what is going on as much as is the statement that the cat is writhing in pain. But the one statement could not be substituted for the other. The statement which includes the concept of writhing says something which no statement of the other sort, however detailed, could approximate to.

I believe the relationship of expert system (or other AI program) rules to the human thought and action which they purport to encapsulate is like the relationship of the cat's writhing to the space-time co-ordinates that describe it. To explain this I will use another analogy — 'record-and-playback' robots: these machines put Winch's imaginary scenario into practice.

Factory production-lines employ robots that mimic human skills in the following way: let us say a paint-sprayer head is fixed to the end of a multi-jointed robot-arm, the movements of which can be monitored and recorded on magnetic tape. The first part of the operation requires a skilled human to guide the spray-head through a job — for example, the complex manoeuvres needed to spray an intricate metal chair. The movements of the spray head are recorded. Henceforward, the spray-head can be made to repeat these movements by replaying the tape so that it drives the motors, exactly reproducing the human operator's movements. Now, imagine that one were to witness such a robot working without understanding the technology involved. One would see the spray-head perform intricately choreographed manoeuvres, reaching round, apparently intelligently, to every nook and cranny in the chair, and switching the spray on and off in just such a way as to cover the intricate shape with the right depth of paint. The robot is transforming a space-time co-ordinate description of the human paint-sprayer's intelligent movements back into the movements themselves. The space-time co-ordinates are a 'behavioural description' of the spraying *action* and the robot's movements are a behavioural reproduction of the original intentional act.[11] So long as the robot continued to spray the same type of chair, there would be nothing to distinguish the behavioural reproduction from the intentional act upon which it was modelled.

The difference between the human and robot paint-sprayer would become obvious, of course, were a chair of a different shape to appear on the production line. The robot would continue to spray in the old way, spraying the paint in all the wrong places, whereas the human would continue as before to perform the act of 'spraying chairs' in a competent fashion. A clever designer could begin to overcome the problem by having the robot trained to spray several different types of chair and arranging for it to be able to 'recognise' which type was

277

currently in front of it and switch to the appropriate section of its program tape.[12] However, every so-far unanticipated type of chair would once more reveal its deficiencies compared with the human. Behavioural reproduction of paint-spraying would only become in principle indistinguishable from human paint-spraying when the pattern for every kind of past and future chair, and the ability to recognise every chair that had ever been invented and every chair that ever could be invented, was programmed into the robot. This, of course, requires an infinite amount of programming.[13] My argument is that the relationship between expertise and the rules and heuristics that knowledge engineers use to represent it is like the relationship between paint-spraying and its record-and-playback reproduction. I propose that we think of such rules as 'the behavioural co-ordinates of skill.' This is what knowledge engineers' rules are.

This chapter makes a less severe distinction between the rules and the actions they represent than Winch makes between the cat and the co-ordinates. Where Winch says that no statement, however detailed, could approximate to the cat's writhing, I have suggested — admittedly as a kind of *reductio ad absurdum* — that an infinite set of behavioural co-ordinates would correspond to action. More importantly, the model makes it possible to see how the extension of a program's rule base allows a machine to mimic action more and more closely. It is easy to see how extension of the rule base makes for increasing interactive competence so long as the user is not looking for trouble. So long as the user is in a charitable frame of mind, the behavioural co-ordinates will be interpreted as representing a range of possible actions on the part of the machine. This interpretation will be identical to the actions imputed to a human whose behavioural co-ordinates of action were the same. Finally, the model suggests that to distinguish a machine with very elaborate behavioural co-ordinates from a human, in a Turing-like test of interactive competence, is not a trivial task. Differences between skill and its behavioural co-ordinates will become increasingly harder to spot unless one knows exactly . what to look for. We must expect to encounter more programs that fool more people more of the time because they do not know how to look for trouble. Perhaps our task, as social scientists, is to work out what trouble comprises, and to teach people how to look for the jagged edges where machines encounter open systems with more variation and creativity than they can cope with. Whether the machines are designed to reproduce manual or intellectual skills, the problem is essentially the same. For sociologists, AI offers us a laboratory for testing our ideas about the distinctions between social action and its external descriptions.[14]

*An earlier version of this paper entitled 'Experts systems and the co-ordinates of action' was presented at the Conference on Technology and Social Change, Centre of Canadian Studies, University of Edinburgh, 12–13 June, 1986, and reproduced in the conference proceedings.

NOTES

1. For recent review articles covering these areas, Collins (1983), Pinch (1987), and Shapin (1982).

2. Some expert-system designers (e.g. Michie 1982) claim that knowledge can be developed autonomously by a process called 'machine induction'. However, the idea of machine induction rests on a very unrealistic model of the scientific process. For an interesting and useful discussion, see Bloomfield (1986).

3. The same problems are discussed, at least implicitly, in some historical literature on skill transfer. See, for example, Harris (1976) and Jeremy (1981). It is also central to the problem of replacement of human skill by computerised machines on the shop floor. See, for example, Jones (1984) and Jones and Wood (1985).

The whole topic also meshes intimately with the 'deskilling debate' (see Berman, 1986). The burden of this chapter is that in the short- and medium-term, expert systems are not likely to have the kind of deskilling effect that a rapid glance would lead one to expect.

4. I am grateful to David Edge for this clipping.

5. For an example of a piece of knowledge that once could not be articulated but became explicit as a result of scientific research, and for some of the complexities in the relationship between different classes of knowledge, see Collins (1987). Finding all of the rules by scientific research is no more possible than the dream of a Laplacean universe. The importance of the rules model is that it allows us to understand the non-solvability of the major problem without being led to the mistaken conclusion that scientific research can find out *nothing* of significance for the problem of AI.

6. For other powerful arguments to show that human action cannot be represented by sets of computer-like rules, see Dreyfus (1979) and Suchman (1985). Dreyfus and Dreyfus (1986) argue that experts do not learn skills in a rule-following way. This is a powerful critique of expert systems, but does not do justice to the social location of expertise. Dreyfus and Dreyfus treat expertise as an individual attribute.

7. The fallacy of the first step is rather like saying that the possibility of reaching the stars is demonstrated by the fact that one can make a start in the right direction by walking upstairs!

8. For a full account of the MYCIN story, see Buchanan and Shortliffe (1984).

9. These figures are taken from a talk given by Karl Wiig at the expert system conference 'ES85', in December 1985.

10. Gilbert claims that irrespective of whether it can succeed in replacing DHSS counter-clerks: '[An] important reason for developing such systems is that they contribute, if only in a small way, to the distribution of information within our society which otherwise would be undemocratically locked into the hands of "experts"' (Gilbert 1985: 2). Whether this ambition is realised must depend on how the systems are implemented and used rather than in immanent properties of their design.

11. For an apt discussion of the distinction between action and behaviour in the context of computers, see Searle (1984).

12. In the first instance we can imagine the robot equipped to 'see' and distinguish between some simple geometric shapes that might be attached to the chairs of different types.

13. Here the programmer would run into the familiar problems of what constitutes a chair — A flat rock? — A sack of straw? Moreover, as Collins (1986) argues with respect to developments in language, an individual programmer cannot anticipate what sorts of objects will come to count as legitimate 'chairs' as culture develops and changes.

14. See Collins (1986) for further discussion of the way to fool machines in the Turing Test.

REFERENCES

Berman, B.J. (1986) Bureaucracy and the computer metaphor. In *Technology and social change*, Centre of Canadian Studies, University of Edinburgh

Bloomfield, B. (1986) Capturing expertise by rule induction. *Knowledge Engineering Review, 1(4),* 30–6

Bloor, D. (1983) *Wittgenstein: a social theory of knowledge*. Macmillan, London

Buchanan, B. and Shortliffe, E. (1984) *Rule-based expert systems*. Addison-Wesley

Collins, H.M. (1983) The sociology of scientific knowledge: studies of contemporary science. *Annual Review of Sociology, 9,* 265–85

—— (1985) *Changing order: replication and induction in scientific practice*. Sage, London

—— (1986) 'The Turing Test: sociological approaches'. Paper presented to the annual conference of The Society for Social Studies of Science, Pittsburgh, 25–28 October, 1986

—— (1987) Expert systems and the science of knowledge. In W.E. Bijker, T.P. Hughes and T.J. Pinch (eds), *New directions in the sociology of technology*, MIT Press, Cambridge, Mass.

———, Green, R.H. and Draper, R.C. (1985) Where's the expertise?: expert systems as a medium of knowledge transfer. In M.J. Merry (ed.), *Expert systems 85*, Cambridge University Press, Cambridge, pp. 323–34

Dreyfus, H. (1979) *What computers can't do*, Harper and Row, New York

——— and Dreyfus, S. (1986) *Mind over machine*. Basil Blackwell, Oxford

Feigenbaum, E.A. and McCorduck, P. (1983) *The Fifth Generation: Artificial Intelligence and Japan's computer challenge to the world*. Michael Joseph, London

Garfinkel, H. (1967) *Studies in ethnomethodology*. Prentice-Hall, Englewood Cliffs, NJ

Gilbert, G.N. (1985) Computer help with welfare benefits. *Computer Bulletin, 1(4)*, 2–4

Harris, J.R. (1976) Skills, coal and British industry in the eighteenth century. *History, 61*, 167–82

Jeremy, D. (1981) *Transatlantic industrial revolution*. MIT Press, Cambridge, Mass.

Jones, B. (1984) The division of labour and the distribution of tacit knowledge in the automation of metal machining. In T. Martin (ed.), *Design of work in automated manufacturing systems*, Pergamon, Oxford

——— and Wood, S. (1985) Tacit skills, division of labour and new technology. *Sociologie du Travail, 4*, 407–22

Leith, P. (1986) Fundamental errors in logic programming. *The Computer Journal, 29(6)*, 545–54

Levy, P. (1987) Modelling cognition: some current issues. In P.E. Morris (ed.), *Modelling cognition*, Wiley, London.

Michie, D. (ed.) (1979) *Expert systems in the micro-electronic age*. Edinburgh University Press, Trowbridge.

——— (1982) *Machine Intelligence*. Gordon and Breach, London

Oldman, D. and Ducker, C. (1985) The non-reproducibility of ethno-methods. In G.N. Gilbert and C. Heath (eds), *Social action and Artificial Intelligence*, Gower Press, Aldershot, pp. 144–59

Pinch, T.J. (1987) The sociology of the scientific community. In G.N. Cantor, J. Christie, M. Hodge and R.C. Olby (eds), *Companion to the history of modern science*, Croom Helm, London

Searle, J. (1984) A walk to Patagonia: 1984 Reith Lecture No. 4. *The Listener*, 29 Nov., 8–11

Shapin, S. (1982) History of science and its sociological reconstructions. *History of Science, XX*, 157–211

Suchman, L. (1985) 'Plans and situated actions: the problem of human-machine communication'. Unpublished thesis, Xerox PARC, Corporate Accession P85-00005

Winch, P. (1958) *The idea of a social science*. Routledge and Kegan Paul, London

Woolgar, S. (1985) Why not a sociology of machines? The case of sociology and Artificial Intelligence. *Sociology, 19*, 557–72

APPENDIX

The following shows what happens when an extra rule is added to
PICKUP. The new listing is as follows.

X will-come-back-to-my-place if
    X has-a-fleeting-smile and
    X will-maintain-prolonged-eye-contact and
    not X is-short-sighted

X has-a-fleeting-smile if
    X smiles-for-less-than-one-second and
    not (X starts-smile-in-less-than-one-tenth-second) and
    not (X stops-smile-in-less-than-one-tenth-second)

A new consultation looks like this:

&. find (X:X will-come-back-to-my-place)

which (X:X smiles-for-less-than-one-second) ?
    Answer is . Jane
    Answer is . Alice
    Answer is . Lucy
    Answer is . Suzy
    Answer is . Carol
    Answer is . end

Is it true that Jane starts-smile-in-less-than-one-tenth-second ? . yes

Is it true that Alice starts-smile-in-less-than-one-tenth-second ? . no

Is it true that Alice stops-smile-in-less-than-one-tenth-second ? . yes

Is it true that Lucy starts-smile-in-less-than-one-tenth-second ? . no

Is it true that Lucy stops-smile-in-less-than-one-tenth-second ? . no

Is is true that Lucy will-maintain-prolonged-eye-contact ? . no

Is it true that Suzy starts-smile-in-less-than-one-tenth-second ? . no

Is it true that Suzy stops-smile-in-less-than-one-tenth-second ? . no

Is is true that Suzy will-maintain-prolonged-eye-contact ? . yes

Is it true that Suzy is-short-sighted ? . no

==> Suzy .

282

Index

active and passive connections 69,
  95, 100
Artificial Intelligence (AI)
  and category errors 17, 33
  and competition 108–10, 112–13,
    145, 148–9
  and hackers 64, 91, 93–4
  and paradigm reactionaries
    xii–xiii, 1–12 *passim*, 80
  and System Dynamics 77–78, 88,
    90, 92, 95
  and the restriction of sociology 64
  attacks on 145–6, 148, 154
    see also relationship with other
    disciplines
  cognitive revolution xiv, 164, 191
  cognitive science 13, 153
  Cognitive Thesis 12–25 *passim*, 46
  commercialisation 149
  communication infrastructure 110
  craft nature 107, 110, 140, 142,
    148
  cybernetics 32–3, 41, 77, 108,
    119–20, 178
  decrease in computing costs 153
  development in UK 119
  development in USA 113
  education 132
  esoteric circle xiii, 74–78, 94, 240
  establishment in UK 137–8, 147
  establishment in USA 115, 138,
    147
  exoteric circle xiii, 71–73
  funding 117
  hardware metaphor xiv, 170, 197
  influence of World War Two 66,
    75, 113, 146, 214–15
  instrumental base 107, 110, 122,
    147
  instrumental reason 62, 89–91
  interdisciplinary nature 145
  intergenerational pattern 110,
    140, 142, 148
  migration of outsiders 143
  monopoly over means of

  orientation xiv, 74, 111, 119
  multigoal characteristic 109
  'neats' and 'scruffies' 74–75
  nepotism 117
  organisational structure 139
  paradigm 11
  prestige hierarchy 144
  relationship with other disciplines
    78–80, 98
  scientific revolution 3
  social construction xi
  social imact/implications of
    61–62, 67–68, 99
  socio-cognitive characteristics
    106–08
  software metaphor xiv, 197
  status of 145, 148
  'strong' thesis 74, 276
  thought collective 71
  thought style 81–88 *passim*
  'weak' thesis 74
Ashby, W. R., 1986–7, 120
Ayer, A., 1, 4, 40

Bateson, G., 41, 178
behavioural co-ordinates of skill 278
behaviourism xiv, 82, 190, 200–3
  see also behavioural co-ordinates
  of skill
Boden, M., 52, 58, 62, 68, 102, 136,
  142, 195–7
Bolter, J. D., 4, 62–3, 82, 102
Boring, E., 189–191, 218
Bruner, J. S., 203–8, 210, 218

Campbell, J., 44, 56
Carnap, R., 173–4, 189, 197, 210,
  218
category mistakes/transgressions 3,
  17–19, 22–3, 26, 33, 55
Chomsky, N., 200–3, 218
cognitive revolution xiv, 165, 191,
  208
  see also software metaphor

283

Printed in the United States
by Baker & Taylor Publisher Services

Printed in the United States
by Baker & Taylor Publisher Services